I0661467

# Animal Dreams

# ANIMAL PUBLICS

*Melissa Boyde, Fiona Probyn-Rapsey & Yvette Watts, Series Editors*

The Animal Publics series publishes new interdisciplinary research in animal studies. Taking inspiration from the varied and changing ways that humans and non-human animals interact, it investigates how animal life becomes public: attended to, listened to, made visible, included, and transformed.

# Animal Dreams

## David Brooks

SYDNEY UNIVERSITY PRESS

First published by Sydney University Press
© David Brooks 2021
© Sydney University Press 2021

**Reproduction and communication for other purposes**

Except as permitted under the Act, no part of this edition may be reproduced, stored in a
retrieval system, or communicated in any form or by any means without prior written
permission. All requests for reproduction or communication should be made to Sydney
University Press at the address below:

Sydney University Press
Fisher Library F03
University of Sydney NSW 2006
AUSTRALIA
sup.info@sydney.edu.au
sydneyuniversitypress.com.au

A catalogue record for this book is available from
the National Library of Australia.

NATIONAL
LIBRARY
OF AUSTRALIA

ISBN 9781743327470 paperback
ISBN 9781743327463 epub
ISBN 9781743327531 mobi
ISBN 9781743327548 pdf

Cover design by Miguel Yamin.
Cover image by Eadweard Muybridge, *Animal Locomotion*, vol. XI, plate 686 (Philadelphia:
University of Pennsylvania, 1887). Via Alamy.

# Contents

# Introduction

DO NON-HUMAN ANIMALS DREAM? People see a dog's paws tremble and move in her sleep, and say that she is dreaming. *I* dream of dogs, but do dogs dream of me? Once, taking lucerne pellets to the sheep in the morning, opening the gate to let them into the paddock, I saw one of them, ignoring the pellets he'd normally immediately devour, race instead at full tilt to a particular tree thirty metres away and examine intently the ground beneath it as if he'd lost his watch or his mother's ring there. But sheep don't wear watches or rings. All I could think was that he'd dreamt of something in that place – something beyond my knowing – and was anxious to find it.

Of course non-human animals dream. This, however, is not a book about them. Instead it's about some of the ways *human* animals, in philosophy, poetry, fiction, and other, more directly political forms, have been dreaming of non-human animals, and about how, very often, for such animals, such dreams have meant nightmares.

In the vast redress we owe the non-human animals to whom we've brought such suffering, we must, accordingly, begin trying to see differently, at every level of our being and in every corner of our culture. I doubt that any one writer or thinker has the capacity to do it all. We can only work on ourselves, at our own levels and in our own corners. I am a writer and have been a teacher of literature, mainly but not exclusively in the English language, and mainly (but not exclusively) Australian. No surprise, then, that a number of these essays – not all – are at work in these areas. Works of undoing, as I've sometimes thought of them.

# Author's Note

WHILE I AGREE IN PRINCIPLE with Jacques Derrida's reservations concerning the term *animal* (see footnote 1 in the chapter 'Animal Dreams'), I find them – as, very clearly, does Derrida himself – impracticable. I use the term variously. Humans, however else we might see ourselves, are also animals. When I am writing of *non*-human animals I try to call them that. There may be instances in this book where I have not done so – that is, have simply used the term *animal* – but in most cases this will be because, in the vicinity or in some other way I feel I have made it clear enough that I am referring to non-human animals not to have to repeat, ad nauseam, that cumbersome term. Elsewhere, if I use the terms *animal* or *the animal* without such qualification, it is because I refer to human *and* non-human animals together, or to that part of their being ('animalness'?) which they share. If I capitalise the word *animal* ('the Animal') it is because I am referring to a (self-conscious) philosophical construct.

I should also say that, in the thought that people might come to this volume to consult specific essays rather than read the collection as a whole, I have tried to ensure that each essay is capable of standing alone, and that there are as a consequence some small repetitions I might otherwise have sought to avoid.

These essays were written over a twelve-year period (2007–19). Some of them were drafted over several years. I have not dated each piece. Dates of any prior publication are available in the Acknowledgements pages.

# The Smoking Vegetarian

*A gaping wound opens before our eyes; we are witnesses of the fact that great evil continues to be perpetuated, and our first task is merely to measure our participation in it. To be objectors in every respect to whatever particular obligation this world attempts to reduce us.*

> – Andre Breton, 'Political Position of Today's Art' (1935)[1]

WHY DO WE STAND IN WONDER before the great paintings, the great sculptures? What is that aura that we seek, that even in the slightest works can capture and enthral us – that emanates from the lines of poetry, that whelms through music, that is part of the atmosphere of novels? Is 'wonder' the best term for it? Critics have wrestled with this question for centuries, resorting to such terms as 'the sacred', 'worship', the 'spiritual', or, rejecting such terms, have found, in that feeling of awe, that yearning, a kind of secular substitute – or evidence that art might *be* a kind of secular substitute.

Certainly it doesn't seem to be rational, this wonder and apprehension, and this, perhaps, is its perpetual appeal. We might,

---

1    *Manifestoes of Surrealism*, trans. R. Seaver and H. Lane (Ann Arbor: University of Michigan Press, 1969), 216.

1

before the work of art, find our mind racing, but the feeling to which I refer is what *sets* the mind racing, not the racing itself: a *stilling*, a confounding, something at the tip of the mind, as a word can be at the tip of the tongue, at once so close and so out of reach that it has others writing of depth psychology, infantile states, mourning for some pre-linguistic plenitude. As if we have found in ourselves a deep hole, a chasm. Even if we choose to ignore it, everything, in this place of art, seems to resonate with its presence.

But that is not really where this essay begins. In a more practical sense it starts in Skadarlija, the oldest part of Belgrade, several years ago. I am in the city at an international writers' festival, one of a handful of Australians there. One of the others, in his early seventies, is a man I've not seen for almost twenty years. In the late 1980s he was a sometime drinking companion when the poets gathered at the university Staff Centre on Friday evenings. After our reading at the National Library of Serbia, an early evening event, we go into Skadarlija, to a restaurant with outdoor tables overlooking the cobbled street. It's a place famed for grilled meats. Our hosts, knowing my wife and I are vegetarian, are keen to make sure there's enough on the menu we can eat. It's clear the rest of the group wants to go there, and it's hard to be spoilers. We assure them there's always something. Salad. Chips. A plain pasta.

It's twilight, late September, warm, the air and the atmosphere delicious. We are all hungry. We order wine, food, pour a glass, drink as we wait. My wife lights up a cigarette. I think about doing likewise but since none of the others appears to be a smoker (not so: after dinner our hosts *all* light up) I decide against it. The man I haven't seen in twenty years seems disgruntled. We make small-talk for a few minutes but then he bursts out with it. 'How can you smoke and be a vegetarian?' he asks, whether of my wife or of me isn't clear. It's a strange question, anyway. Not only am I unable to see any inherent connection between smoking and vegetarianism, but I also remember him, all those years ago, being a particularly heavy smoker. He's given it up, evidently.

We attempt to answer, tell him, though he seems to have trouble believing it, that we're not vegetarians for our own health, but out of concern for non-human animals (we're dissembling already: we're not vegetarians, but vegans, but that's even more contentious). He continues, rude and intrusive, long after my wife's one cigarette has been stubbed. It's not clear what bothers him most – that we are smokers, or that we are vegetarians. It's as if he feels he's caught the whiff of some deep hypocrisy and is determined to ferret it out. The evening begins to turn sour. It seems as if nothing will stop him.

Vegetarians are never popular.[2] Recent converts especially. Vegans far less. It's OK if it's for one's health – indeed one can receive a measure of understanding and sympathy, as presumably one might if it were for religious reasons – but if it's an aversion to animal slaughter, animal cruelty, a *con*version, it unsettles people. One's friends can't stop talking about it, and then most likely complain that it's you who's become obsessive. As if they feel betrayed. Unobtrusive as you try to be, you've become a walking chastisement. Some people can get very vehement indeed. 'Vegetarians, and their Hezbollah-like splinter faction, the vegans,' declares celebrity chef Anthony Bourdain (*Kitchen Confidential* 70), 'are the enemy of everything good and decent in the human spirit.' 'There is nothing new about turning vegetarians into figures of fun,' writes Colin Spencer (*The Heretic's Feast*):

> disciples of Pythagoras became stock characters in Attic comedy ... But other societies felt such criticism was no laughing matter and the outsiders were reviled. Vegetarians then became criminalised and were considered blasphemers and heretics. (xiii)

The Catholic Church, for example, has always been in two minds. There are vegetarian orders, even vegetarian saints, but the Church has historically been at pains to point out that vegetarianism is common to most of the major heretical movements it has tried to repress. It's an aspect of Manicheanism. The Bogamils were vegetarian. The Cathars. The Albigensians. Make a case for the inherent vegetarianism of Christianity (as some of these movements did) and you'd have to shoulder a huge weight of contradictory discourse. Christ is the *lamb* of God. 'This is my flesh,' he says. 'Take, eat, in memory of me.' Christianity, arguably, is organised – like the religion from which it

---

2   True enough at that time (2005) and the time this essay was written (2008), though much less so now.

evolved – around eating rituals, proscriptions of the consuming of this or that kind of flesh (long lists in Deuteronomy and Leviticus) that are at the same time encouragements to eat other kinds, to *sacrifice and eat other kinds* (horror and shame displaced: the making sacred as a mode of abjection). Indeed the Judeo-Christian tradition, with its consistent encouragement to eat meat, its denial of souls to non-human animals, its concentration upon Man as the pinnacle of creation, has done much to bring about the culture of animal cruelty that flourishes today. Genesis 9 spells it out emphatically:

> And the fear of you and the dread of you shall be upon every beast of the earth, and upon every fowl of the air, and upon all that moveth upon the earth, and upon all the fishes of the sea; into your hand are they delivered. Every moving thing that liveth shall be meat for you.

The Split, I call this, the Scission, by which the human, upon promise of eternal reward, is led to deny the core of its being.

To a vegan, vegetarianism and such paraforms as 'pescatarianism' seem half-hearted. The less meat one eats the better, yes, whatever the reason: it can't be good for one's overall anxiety level, subliminal or otherwise, to be consuming flesh still flushed with the terror of imminent execution ('humane' slaughter? think again). Avoidance of meat is also good for one's weight, blood pressure, cholesterol level, predisposition to diabetes, etc., to say nothing of one's own feeling about oneself and one's relation to the world, although these latter will be factors only if one has changed diet out of compassion for the animals one might otherwise be eating, and it's here that half-measures become less and less plausible. I won't embark upon the cruelties of battery farming or the dairy industry. There are many ways to inform oneself of such. Suffice to say that, if one is determined to minimise one's impact upon the other animals we live amongst, then one's diet will eventually be free of eggs and of dairy products.[3]

---

3    A 'spent' cow – one that no longer gives sufficient milk – or 'spent' hen will be slaughtered anyway. Arguably all one has done, as a vegetarian, is to support the addition of a prolonged period of prior suffering.

And, of course, if that *is* one's concern, one doesn't stop there. One avoids leather, fur, wool. One avoids products that contain ingredients that come from animals, and products tested upon animals. One avoids *feeding* animals *to* animals. And initially this regime will seem difficult, for one finds such products everywhere. They are components of the furniture we use; they are in our cosmetics, medications, soaps, shampoos; most clothes contain them; they help to 'clear' most of the wines we drink; they are there even in the ink and spines of the books we consult in order to learn about them – so omnipresent, indeed, that one could speak of very nearly the entire material world of humans as deeply intertwined with and supported by the world of non-human animals, except that 'supported' is woefully misleading: better to say 'dependent upon the cruel treatment and the killing of animals' – riding, as it were, on a tide of suffering.

Is *that*, then, what this essay is about? Yes, and no. My account of the absolute pervasiveness of the animal in human *material* culture (i.e. of the *non*-human animal in *human*-animal culture) is to preface the point that the suffering, subjugation and debasement of such animals are just as pervasive in the world of thought as in the world of things – a pervasion the apprehension of which, hard enough in the first place, is rendered all the more difficult firstly because it is a matter less of active *thought-against*, as in overt proscriptions of non-human animals or injunctions to animal cruelty, than it is of *blindness to* them, brought about by the overwhelming orientation of attention elsewhere (i.e. onto the perpetuation and 'advancement' of the human); secondly because it is, therefore, a pervasive *absence* rather than a pervasive *presence*; and thirdly because, as with all problems-of-thought of this nature, perception of the extent of this pervasion must be done *with a mind already pervaded*.

Although one might want to preface such a statement with the assertion that such divine powers were in the first place invented in order to issue just such an injunction (so much of the sacred having been made by humans for human purposes), the assumption of dominion over the animate world was never so clear and simple a matter as (as the Book of Genesis presents it) an injunction from God. It's a far more ancient process than anything the Bible might legend, a key factor in the genesis of culture per se. A shift, let's say, from hunting for survival, animal amongst animals, to the cultivation of meat alongside one's grain. A set of mental procedures, eventually, consolidating into rituals, and rituals-of-*thought*, to enable one to deal with and rationalise the killing of creatures that have become part of one's immediate environment and that, if not necessarily loved, one has at least observed the co-creatureness of and recognised something of one's own creatureness in. It is harder, a kind of fratricide, to kill that-which-one-*is* than to kill that-which-one-is-*not*, and so the assumption of this aforesaid dominion entails a denial of that co-creatureness, an emphasis upon difference rather than sameness, and a machinery of thought that exaggerates and consolidates, indeed largely *invents*, that difference. Metaphysics surely has some part of its origins in this, a shift, in this regard, into bad faith, inauthenticity, a reinvention of oneself as something other than what one is, a denial of one's animal being, a dividing of oneself against oneself, a suppression so deep that it might best be described as a *wound* that we must carry, must attempt to deal with, have been conditioned, ironically, to turn to metaphysics to attempt to soothe and explain, without ever really knowing what it is, or where it comes from.

What do we do to ourselves when we slaughter? What do we do to ourselves when we eat our own kind (apart from drawing death so deeply into ourselves, into every cell of the mind as well as the body [as if these can be distinguished])? Deny our kindredness, reinvent ourselves, attach ourselves to another place, give ourselves wings as we might, the fact remains that we also brutalise ourselves – that we also know that we *are* animal, *are* kindred, and *have* embarked upon a kind of cannibalism – a long, slow holocaust – and that repressing this and the shame and guilt that attend it – articulating a whole metaphysical complex to give it a specious validation (a complex akin to the way we process, package and describe our meat to disguise its origins) – we have created a deep schism in ourselves. This schism, this Wound, albeit so deeply repressed that we can only begin to guess which of and in what manner our beliefs and institutions are its proxies, is one of the givens of our culture.

This process is not only one of the main currents in the development of human thought, but is alive in each of us, a complex navigated in the enculturation of every individual. We don't, by and large, want our children to see the slaughter that brings meat to their table – indeed it so unsettles *us* that we have developed complicated systems to mask it – but children seem particularly inclined to identify, and to express their horror at, such processes. But then, of course, we see them harden into an adulthood in which, rather than outgrow or overcome it, they in some parts suppress, in some parts sublimate, their horror, compounded now by a sense of shame, of having betrayed.

Any such position will face familiar arguments. Most common is that, as the human is biologically constructed as omnivore, it is 'natural' to eat meat. This point is simplistic – it's far more complicated than that – and largely irrelevant. People who make this point tend to assume that human ancestors have been meat-eaters but there's a growing field of evidence that our ancestors were mainly herbivore. The shape of our teeth may allow us to eat meat, yes (though they're a bit flat for the job), and our digestive system *can* handle the flesh of others, but it's also true our bodies bear organs that we no longer use, and that the presence of a capacity is no obligation to exploit it. We also, as it happens, have the capacity to choose. There is, for most of humankind, no longer any necessity to depend for sustenance upon the slaughter of other creatures. Almost as common is the argument that meat is necessary for a balanced diet. This, while fervently promoted by the meat industry, is balderdash.

A third argument, Protean in its manifestations, is of a different kind. An anecdote should illustrate it clearly enough. A Greek friend, describing Easter in his native village, speaks of the killing and roasting of the lamb ('Take. Eat …'), and the way this ritual becomes a kind of initiation for the children. He too, he tells me, is troubled by the tiers of pre-packaged animal parts on supermarket shelves. The modern urbanite has lost touch with the land, he says, and with ritual. Ideally one eats only meat that comes from animals one kills oneself, so that one knows where the meat comes from and what has been done to bring it to the table. Some sense of *taking responsibility* is involved, a notion that the killing is somehow acceptable if one does it oneself, does not delegate it or expect that it not intrude upon one's consciousness, although this, on the other hand, seems to admit that there is a weight, a problem, to be taken responsibility *for*, and that this killing should not be an ordinary or easy thing to do.

The word ' sacred' has been in the air for some time as we talk and now, at last, appears, though it's not clear whether what is sacred is the life taken or – as if the one confers its sacredness on the other – the taking of the life. I think – more sceptically than I might once have done (I can think of no reason why 'tradition', 'belief', etc. should absolve one culture more than another) – of the native American tribes who thanked the spirit of the buffalo they hunted, or of the Inuit who offered fresh water to a seal they had killed in order to appease his or her spirit. Yes, life *is* sacred, if I can employ that term in a secular sense – there are many things that are sacred in such a sense – but I am uneasy about the term, firstly because it *is* so definitively *non*-secular, so much within the economy of the religious, and secondly because it is so subject to critical misuse.

The term 'sacred' is itself sacred, it seems, and I'm not sure that it should be. All too often its invocation is a sign that something ethically questionable is being placed safely beyond discussion. It also *objectifies* the thing it qualifies – separates and segregates it. We began to need God, the sacred, etc., as separate entities, when we removed something from ourselves, or sought to remove ourselves from where and what we had been: when, with an act – a long, slow act, albeit – of supreme arrogance, we said that what was *here* (the non-human animal, but not only that) was about *us*, was *not* sacred, and could therefore be subjected *to* us.

As to the non-secularity of the term, I have hypothesised already that metaphysics were born out of the need to establish the human as something above or beyond the animal, as a means of asserting and consolidating that *dominion* which the canny makers of the Bible needed to establish before any other human attribute could be presented. But any implication – and it *is* my implication – that such a development is universal, rather than the product of a particular culture, must confront the existence of *vegetarian* religions, past and present, that argue the sacredness of all life and proscribe the taking of it in any form. My difficulty with these – mitigated by gratitude for *any* abatement of cruelty – stems from the way that, even in their apparently contrary direction, such religions still have as their basic premise the progressive *extraction* of the human from its animal predicament. If they proscribe the eating of flesh or the taking of life, they do so on the understanding that this is in the individual's best interests. The compassion they advocate is still, as it were, framed as a compassion-under-duress rather than voluntary and a logical product of one's being.

Another anecdote. I am walking through the university on the first day of first semester. The place is teeming with new students looking for their first lectures, for coffee, for the library, for their friends. I am struck as always by their confusion, their wide-eyed, *calf*-like innocence. How many are vegetarian? How many vegan? One in a hundred? Less?[4] I imagine them carrying their sandwiches for lunch – ham, chicken, cheese – or ordering at one or another of the cafeterias: prawn laksa, beef curry, hamburgers. Even if they wished to think differently, how could they? How is this place – this *university* – going to help them? I try to think of a discipline that is not in some way dependent upon dominion and the exploitation of non-human animals, but which would it be? Such disciplines surely exist, but for the moment I can't think of them. Mathematics? Physics? Possibly, but I can think of few others. Certainly not Philosophy, where every major figure, meat-eater or otherwise, either oils the sympathetic machinery of metaphysics or prefaces their work on advancement of the human. Certainly not Literature, my own discipline, where the entire pastoral tradition turns about an unspoken centre of cruelty; where the greatest novels rely as much upon dinner parties and the carving of beef as they do upon battles and the intensities of human emotion. Certainly not Law, Medicine, Education, Psychology. And certainly not Art or History. The university – my beloved turning (and disappearing) world of knowledge – seems suddenly, to my absolute dismay, but another huge and complicated agency for the unreflective propagation of cruelty, swallowing these students as an abattoir swallows its victims.

Overstatement? Yes, of course, but a reeling vertigo is not unusual before one gets one's bearings in this new place-of-thought. Reorientation towards non-human animals demands a deep and extensive *de-centring of the human*. One questions directly the right of

---

4    In the years since this essay was written (2008) veganism has become
     something of a fashion. We might now, on some campuses, be speaking of
     one in ten.

*Homo sapiens* to be the point of everything, proposes that the human be *amongst*, not *above*, that it be *one of*, not *the*. Hitherto, one realises, all understandings of value, all standards of good and evil, had been predicated upon the continued dominance of the species. But what if that were *set aside*?

If one has not been brought up to it, this perception can come as a dramatic destabilisation. It can seem, for a time, as if all the edifices of one's previous understanding, their centre-pin gone, list, threaten to topple, their contingency exposed. The noblest expressions of human nature and purpose can appear blinkered and naïf. One can be shocked at what one finds oneself reconsidering. Surrounded by people brought to a veritable standstill by the insurmountable difficulties of their own existence, for example, one can find oneself thinking that the melancholia that seems the keynote of our time is not much more than an immense, endemic Narcissism. 'How can you talk about the suffering of animals,' I'm asked, 'when there is so much *human* suffering?', and while one of my answers is clear, that compassion is self-replenishing and without borders, and that there is no need – especially when most of us give so little to anything other than our own immediate interests – to apportion it, I am all too aware that what is given towards the saving or rehabilitation of a human life, while it must, yes, *be* given, is nevertheless almost certainly also extending the life of one who will continue to eat meat, continue to contribute to the suffering of animals.

A dilemma? Or merely a paradox. It is hard to know. A recent conversation with a friend turned to matters of animal cruelty, then inevitably to veganism:

'So you don't eat *any* meat?'
'No.'
'Fish?'
'No.'
'No meat, no seafood – and eggs? milk products?'
'No.'
'And this is not for your health but to avoid cruelty.'
'Yes.'
'But where does it stop?'

We talk, then, about 'sentient' creatures. How does one define sentience by anything other than a *human* understanding of the concept? And how can one say that it is unacceptable to kill 'sentient' creatures but alright to kill others? Aren't plants also in some measure sentient? How can we say that they are not?

Vegans often go further, for just these reasons. Some become fructarians, eating only those parts of the plants that the plants produce *to* be eaten as part of their own process of reproduction. Some argue that we should only eat *wild* fruits, regarding cultivation itself as a kind of exploitation. My friend and I agree that, following a strict logic in this regard, one might end up living on air, scarcely able to move for fear of hurting some living thing in the process – effacing oneself in order to avoid cruelty to other creatures, while those other creatures show no such compunction. An absurdity, of course, and yet also not. A paradox. A kind of point zero. And yet one cannot – this is surely, like *argumentum ad hominem*, one of the great errors of thought – allow the difficulty and possible absurdity of the extreme to undermine the principle. It is only in the realm of thought-by-itself – thought alone – that thought can be tidy and without paradox. Thought-in-the-body is a different thing.

The conversation tails off. Driving away, I find myself thinking firstly of the *logical difficulty* (i.e. that it is *logical* that there *would* be difficulty) of *de-centring the human* in this way. All of our thought – our Logos – pivots about that axis: the *machinery* of our logic, our language, its grammars, its systems of metaphor. How could one expect anything other than paradox, embarrassment – a kind of radical and systemic *stupidity* – in one's attempt to think into so radically different a place? How *can* one think, against the core of thought? And then, those words 'point zero' still ringing in my ears, realising that what is (now) needed is a Descartes-like return to a kind of point zero of thought, a starting from scratch, with compassion, rather than wound-spurred anxiety, as one's foundational premise.

I am back at the Wound again. As my opening intimated, I want to speak of 'the animal' in literature and art in a sense different from any I might so far have adduced. Neither in terms of its presence, let's say, nor of its absence, but of something as likely manifest in the curve of an archway or the shadow of a chair. Once, at an exhibition of the paintings of Giorgio Morandi, I saw a man in a business suit standing before a still life, weeping. The painting represented a set of glass bottles on a tabletop. Four glass bottles, and an open canister, nothing more. Can it be that people go to art galleries in order to sublimate their pain? Or to seek its reflection, its echo? Could it be that, in each work they pass, the Wound searches for itself (or for balm, for letting)? Albeit largely unbeknownst to them, just as the artists themselves cannot have known – at least, not many of them – the nature or origin of that intensity they wished to capture or restore, the idea they wished to return to its body? The Church and the Gallery closely linked, in a complicity that almost passes understanding? But that may be the point, to *pass* understanding: to be a place – places – to keep from the mind what the mind cannot bear.

I've given up smoking, as it happens, or rather smoking has abandoned me. Not for my health, although I do feel the better for it. A bit like veganism itself. One mightn't do it for one's health, but that might improve nonetheless, and not just the physical. The Wound is so deep that there's probably no fixing it, soothe it as we might, but with the de-centring of the human there can also be (how to put this?) *a de-centring of the self in one's own life*, a shifting of attention away from the needs and wounds of the self – those *lacks* that are, paradoxically, one of the mainstays (the matrices) of contemporary society – towards the needs and wounds of the other animals we live amongst, in such a way as amounts, ultimately, to a powerful and empowering redefinition of the self, a new and radical kind of wellbeing.

# Cracks in the Fray

## Re-reading 'The Man from Snowy River'

WHEN I WAS EIGHT YEARS OLD, in Canberra, my father took me to
the Institute of Anatomy where, amongst the vast array of pickled
foetuses and body parts (brains, stomachs, lungs, hands with six digits)
preserved in jars with labels in immaculate copperplate, I found myself
before a huge glass tank – at least, to an eight-year-old it was huge –
staring at a pale, pasty lump about twice the size of an Australian Rules
football. Obviously it was an object of considerable importance since,
instead of the copperplate description, this (in my distant memory) had
a large wooden plaque upon which, in gold letters, were the words 'Phar
Lap's Heart'.

I had no idea who or what Phar Lap was. My father explained
that Phar Lap had been a famous racehorse. It was another six years
before I found myself at the Victorian Museum, again with my father,
staring up at a scruffy and exhausted Phar Lap himself, and I don't
know how many more before I took in something of the story of how
Phar Lap, a gelding (i.e. castrated so that he'd not be distracted by other
urges), had won the Melbourne Cup and so many other of his races,
and how he'd been poisoned by parties unknown after winning a major
race in Mexico. It would still be a little time yet before I made the sad
connection (but it is there to be made) between Phar Lap – whose heart
is in Canberra, whose skin is in Melbourne, and whose bones are in
New Zealand – and Truganini.

Phar Lap, it seems, was a much-loved horse, though one has to wonder whether it was the horse himself who was loved or the fact that he made money for so many. Australians have always loved their horses, or rather their horse races. Where else in the world does a horse race stop the country? And so love a winner that they'll happily claim winners who are not exactly their own: Phar Lap (the name, part Zhuang, means 'Lightning') was in fact a New Zealander.

Andrew Barton Paterson (1864–1941) loved his horses, and loved horse racing (even his nickname, 'Banjo', came from a horse), and not just as a spectator. A rider from his earliest childhood, he not only wrote many poems about horses and horse racing – one of them perhaps Australia's best-loved poem – but was apparently a fine polo player and successful steeplechase rider.

Australia, it's often said, is a land of immigrants. Mostly. It's not often pointed out that a great many of these immigrants are non-human animals, but of course such animal immigrants were there from the first moments of the invasion. The First Fleet not only brought convicts and their gaolers, but also pigs, sheep, goats, cattle, chickens, etc. (Norwegian rats, German cockroaches …) – and horses. And while none of these had in any official sense been convicted, it's fair to presume that not many of them (we might have to except the rats and cockroaches) came of their own free will. Indeed it can sometimes seem as if Paterson himself is quite consciously making this connection, giving as he does to some of the key horses in his poems convict-oriented names like 'Pardon', 'Reprieve', 'Regret' (if that is indeed a horse) and 'Manumission', although I admit readily that this habit might instead be a sign of his times (he was four years old when transportation came to an end: i.e. it was still very much in the air), or of the fact that he was a lawyer, or (not so unlikely, for a gambler) an attempt to register some more philosophical reflections upon the nature of earthly existence.

Other animal immigrants (rabbits, foxes, Indian mynahs) came as the colonies established themselves and some of their more successful human animals endeavoured to reflect lifestyles of their native lands. Still others (camels, cane toads) were brought in to work in agriculture or construction. And for many of these non-human immigrants, and probably before it was so for many of the humans, Australia turned out to be a land of plenty, particularly for those who were lucky enough to escape or be turned loose. Two hundred and twenty years after the white invasion there are, in their various parts of the country, well-established tribes of rabbits, goats, pigs, camels, horses, foxes, toads. In most cases they are considered pests, a term reserved for creatures that the human population is unable or chooses not to commercially exploit (or needs an excuse *to* exploit), or that get in the way of their exploitation of others: sheep and cattle, for example, are not considered pests, although their hooves have destroyed so much of the oh so friable surface of the country. And it is reasonable to assume that some of these wild tribes established themselves quite early. Certainly the tribes of horses.

By the time Paterson was born, in 1864 on Narrambla station near Orange, New South Wales, 'brumbies', his 'wild bush horses', were a significant 'problem' in the area. Although the particular nature of this problem – whether it was a matter of competition for feed, a fear that bloodlines of enslaved horses would be contaminated by wild intruders, or a difficulty in maintaining one's fences – is not quite so clear (particularly as, captured and broken in, or simply slaughtered for meat and skins, they represented what one might have thought of as a valuable resource), there were at this time earnest enough attempts to eradicate them. That is where, arguably, 'The Man from Snowy River' begins.

I've said that Paterson was a lover of horses and of horse racing, but, and although in one sense I don't doubt it at all, I was merely relaying a cliché. In truth Paterson's own poetry demonstrates most eloquently how hard it is to see how someone could love horses and horse racing at the same time (and even though there can be no doubt that this oxymoronic state of emotion – one struggles to find a word for the depth of its paradox – is and always has been something like the lifeblood of the racing industry). Paterson may 'love' horses but, in the interests of a good horse race, he seems most ready to see them treated with nothing less than savagery by their owners and riders, and indeed to celebrate the fact.

See, for example, 'The Amateur Rider', in which a young man (in a scenario very similar to that of 'The Man from Snowy River') who fails quite utterly to impress with his attire –

Ride! Don't tell me he can ride. With his pants just as loose as
  balloons,
How can he sit on a horse? and his spurs like a pair of harpoons

– very quickly changes the speaker's tune when it comes to the ride
itself:

Sit down and ride for your life now! Oh, good, that's the style –
  come away!
Rataplan's certain to beat you, unless you can give him the slip;
Sit down and rub in the whalebone – now give him the spurs and
  the whip!

Or 'Old Pardon the Son of Reprieve', in which outsiders, taking their
horse to the Menindee races, are tricked by the locals who endeavour,
the night before, to ensure the horse cannot run:

They got to his stall – it is sinful
  To think what such villains will do –
And they gave him a regular skinful
  Of barley – green barley – to chew.

He munched it all night, and we found him
  Next morning as full as a hog –
The girths wouldn't nearly meet round him;
  He looked like an overfed frog.
We saw we were done like a dinner –
  The odds were a thousand to one
Against Pardon turning up winner,
  'Twas cruel to ask him to run.

But of course they 'ask' nonetheless, and although (these were 'heat'
races: best results out of three) 'all lathered and dripping with sweat'
and clearly very ill after the first, he comes good in the second ('under
the whip') and romps home in the third.

Or see this, from 'The Open Steeplechase' (wherein 'clout' refers to the horses' hooves and/or fetlocks striking the top of the barrier they are attempting to jump across):

> But the pace was so terrific that they soon ran out their tether –
> They were rolling in their gallop, they were fairly blown and beat –
> But they both were game as pebbles – neither one would show
> the feather.
> As we rushed them at the fences, and they cleared them both
> together,
> Nearly every time they clouted, but they somehow kept their feet.
>
> Then the last jump rose before us, and they faced it game as ever –
> We were both at spur and whipcord, fetching blood at every
> bound –
> And above the people's cheering and the cries of 'Ace' and
> 'Quiver',
> I could hear the trainer shouting, 'One more run for Snowy
> River'.
> Then we struck the jump together and came smashing to the
> ground.

Seemingly Paterson is unaware of the paradox here – so clear that it could be described as an immediate contradiction – that, if it were *true* (and how could it be?) that 'neither one' of these horses 'would show the feather' (itself an intriguing metaphor to explore), then it would be most unlikely they'd have to be so spurred and whipped that they were literally spattering blood as they galloped. But this is sound – and gallop – above sense, and sentiment above consciousness.

It's precisely at such points of crisis – moments of clash between orders of thought and sentiment that threaten (to use a more contemporary metaphor) to crash the system of the poem – that, as if to try to slip out sideways, the horses are most emphatically personified, if only so that they can then be claimed to share that strange human propensity to see blood and injury in the interest of sport as badges of honour and of courage (this happens, as we shall see, most poignantly in the conclusion to 'The Man from Snowy River').

Love of horses *and* love of horse racing? If one loved one's children in such a way one would end up in prison. It might be noted that the steeplechase is in more contemporary parlance called 'jumps racing' and that, even as this essay is being written, the State of Victoria, heartland of Australian horse racing, is being pressured to ban jumps racing after a few too many images have reached the wider public of horses killed directly on the track or being 'put down' after being injured (most often through 'clouting') during a race.

But Paterson, it would seem, is a man lost amongst his contradictions, just as a nation which idolises his poetry – and there is no doubt whatsoever of his skill as a balladeer – might be thought to be a little lost amongst its own. To look at this from a different angle, and approach another contradiction, consider 'Only a Jockey', in which Paterson treats, rather unconvincingly ('Ere the first gleam of the sungod's returning light'?), the sad death of a fourteen-year-old jockey at training, and castigates, as a flagrant example of human greed and concern for one's pocket, the newspaper report of the incident for having said 'The horse is luckily uninjured', failing quite utterly to confront the fact that it is not some vague inhumanity (the 'they' we so like to blame things on) that has brought about this death, but the industry Paterson himself so loves and is so much a part of.

JONATHAN SWIFT, in Book IV of *Gulliver's Travels*, famously reversed the tables on horse/human relations and made the horses the masters. I'd like to try something similar in a reading of 'The Man from Snowy River': not so much to make horses the masters, tempting as that might be, as to return to them, *Rosencrantz-and-Guildenstern-are-Dead*-like, something of the narrative presence and agency the poem itself denies them.

The poem is about brumbies. Although it's possible the word wasn't in such wide currency at the time, and although Paterson doesn't, here, call them that, the term repays some attention. It refers, of course, to horses, and to the descendants of horses, who have run wild: one could say horses who, either because they've managed to free themselves or because they've been turned loose by people who can no longer afford to keep them, have escaped human servitude and are fending for themselves, organising their own tribes, their own nomadic communities.

No one really knows where the term 'brumby' comes from. One likely source would seem to be a Sergeant James Brumby who, in 1804, left some horses to fend for themselves when he quit his property at Mulgrave Place, on the Hawkesbury, to move to Tasmania, but there's also a suggestion that the term comes from the Pitjara of southern Queensland, in whose language 'baroomby' means 'wild'. A group of brumbies, it's also worth noting, is called a 'mob'.

Of course, there are many who would say the poem isn't about brumbies really (and we'd probably find almost unanimous agreement that it is not in any way about Indigenous Australia!), but about a *man* from the Snowy River country, and his ride, albeit a ride on a remarkable horse. Numerous critics and commentators have assembled a *dramatis personae* for the poem – there is Clancy of the Overflow, there is Harrison ('who made his pile when Pardon won the cup'), there's an 'old man' who in one instance may be that Harrison and in another the owner of the runaway horse, and there's the 'stripling' ('stripling' in the poem, 'man' in the title) whose remarkable riding becomes the centrepiece – but for my purposes it would be more useful to assemble a *dramatis equi*. The poem in fact features a number of horses, present or alluded to, and it pays to give them some thought (at least, it *would* pay, if compassion and awareness of past cruelty were of greater general concern than they are).

There is, most immediately, 'the colt' whose escape has occasioned the chase and so the poem. We don't know how or why he's escaped – whether, say, someone has left the stall unbarred, or whether he's broken his way out – but clearly he's not wanted to stay at 'the station', and desires to join the relatively free company of his own kind. Perhaps he's sensed the presence nearby of his wild brethren (we might once have referred to this, *pace* Jack London, as 'the call of the wild'); perhaps it's just that he's found them once he's found himself free. In either case he's too valuable to his human owner to be allowed to get away with this.

Next there is 'old Regret', most likely his dam, which is to say mother, though there's a possibility (it's customary, when giving a horse's pedigree, to say 'out of' rather than 'from' ['the colt from old Regret'], though 'out of' would not scan so smoothly, and neither is used for the sire, only the dam) that 'Regret' is the name of a location – a station – rather than a horse. More than one student has come to me, when I've lectured on this poem, with the 'discovery' that Regret was indeed a famous racehorse – winner, amongst other events, of the 1915 Kentucky Derby – but although she did in fact have several foals (eleven! probably not voluntarily), and although these foals would have been very valuable, the dates rule her out: she was born in 1912, raced in the years of the Great War, and died in 1934. The poem was written in 1890. The latest that *Paterson's* Regret, if he/she had raced at all, could have done so would have been in the mid-1880s.

Then (passing over, for the moment, 'the wild bush horses') there is Pardon, who appears to have come directly from Paterson's own childhood, though Paterson himself seems to have been in two minds about some of the details. In the *Sydney Morning Herald* in 1938, in a series entitled '"Banjo" Paterson Tells His Own Story', we read of him being taken, at the age of eight, by the station roustabout, to the Bogolong Town Races. While there, he sees 'a Murrumbidgee mountaineer about seven feet high' making off with the undersized saddle from his (Banjo's) pony and putting it on a racehorse:

> Running over to him, I managed to gasp out: 'That's my saddle.'
> 'Right-oh, son,' he said. 'I won't hurt it. It's just the very thing the doctor ordered. It's ketch weights, and this is the lightest saddle here, so I took it before anybody else got it. This is Pardon,' he went on, 'and after he wins his heat you come to me an I'll stand you a bottle of ginger beer.'

Pardon won the race – not a 'Cup' in this instance, but the Bogolong Town Plate – and the eight-year-old got his reward. In a *Sydney Mail* of later the same year (1938) the story has changed somewhat, though not in a way that cannot be reconciled with the former version: Pardon is now owned by Paterson's uncle, and (as in 'Old Pardon, the Son of Reprieve') breaks out of his stall into a bale of lucerne the night before his run (in 'Old Pardon, the Son of Reprieve', it might be noted, the race is named 'the President's Cup').

The fourth horse to be specifically mentioned in the poem is the stripling's, in the description of whom there is a measure of tension, since Paterson is at pains to establish at once his unpromising and unlikely appearance and his remarkable character. He is, that is, 'a small and weedy beast', but also

> something like a racehorse undersized,
> With a touch of Timor pony – three parts thoroughbred at least –
> And such as are by mountain horsemen prized.
> He was hard and tough and wiry – just the sort that won't say die –
> There was courage in his quick impatient tread;
> And he bore the badge of gameness in his bright and fiery eye,
> And the proud and lofty carriage of his head.

– although, a slight tension here also, the 'old man' (no matter whether Harrison or the owner of the colt: each, surely, should know his horses better) doesn't see it:

> But still so slight and weedy, one would doubt his power to stay,
> And the old man said, 'That horse will never do
> For a long and tiring gallop – lad, you'd better stop away,
> Those hills are far too rough for such as you.'

But maybe it's the stripling, not the horse, whom they're concerned about, trying to protect him, or just get rid of him, by suggesting his horse – not he – isn't up to it. There's a subtle but interesting eliding of the species barrier here, just as there is more evidently in the coming parallel between the stripling and the runaway colt. (The stripling tries, very successfully, to prove himself a rider among riders by in effect riding down himself.)

And then, of course, there are the 'wild bush horses'. If the overt plot-line in the poem is the courageous and extraordinary feat of the stripling's ride, and another, supposedly, the recapture of the valuable colt (just broken *out*), the covert plot-line, just as strong, is the dragging into captivity – Kubrick's *Spartacus* comes to mind – of a whole tribe, made up of escapees and emancipists and the descendants of such, that has for a long time roamed free, now to be either *broken in* or (but at this point in time this is thankfully less likely) slaughtered. They are remarkable horses; the poem has told us that – hardy, almost impossible to catch, perhaps themselves with a touch of thoroughbred and Timor pony – yet the stripling proves their match:

> And he ran them single handed till their sides were white with
>     foam,
>     He followed like a bloodhound on their track,
> Till they halted cowed and beaten; then he turned their heads for
>     home,
>     And alone and unassisted brought them back.

From a non-anthropocentric perspective that last line is an outright lie. The stripling alone and therefore on foot could have done nothing except stand and stare at the disappearing mob. The wild bush horses have been ridden down, not just by the stripling, but also and *principally* by his horse, albeit under considerable duress, and it might pay to consider the latter's predicament.

The next lines, which turn towards him, are some of the most troubling in Paterson's oeuvre:

> But his hardy mountain pony he could scarcely raise a trot,
>     He was blood from hip to shoulder from the spur;
> But his pluck was still undaunted, and his courage fiery hot,
>     For never yet was mountain horse a cur.

The driving rhythm and the general sound-over-sense sentimentality might have us coasting over the top of these words, but if we resist, and instead draw into the mind's eye the scene presented, we have a horse who has been so gouged by the spurs of his rider that he is – the lines *say* it – *covered* with blood, from shoulder to hip. We might say that this is merely expression, epic exaggeration as befits the general quasi-Homeric frame, but the accuracy of the detail here – the actual *experience* (Paterson's) reflected in the lines – should give us pause. The spurs would dig in to the horse's hips as he was ridden up-hill, and into the shoulders as he was ridden down. The extent of his injuries – of his deliberately inflicted wounds, for they are that – is quite in accord with the task he has been forced to perform. Paterson, in this one sense, is quite well aware of the likely wounds inflicted upon a horse by a ride of this kind.

Having just described the horse as blood-drenched, barely able to raise a trot, the poet's too-obvious bench-press attempt to lift the moment back into the heroic ('But his pluck was still undaunted, and his courage fiery hot') is as impressive as it is ludicrous (pluck? courage? at spur-point? what choice did he have? I suppose he could have resisted, thrown his rider, joined the wild horses himself, but then where would the poem be?), and the line which follows, while one is inclined to think it quite true on the evidence so far, also brings about one of the poem's more blatant contradictions, for haven't we been told just five lines earlier that the 'wild mountain horses' are 'cowed and beaten'? (An animal-centred reading of the poem [and this reading is only partly so, a gesture] would find one of those words particularly indicative.)

But this is to speak of only a part of this horse's predicament, which from a certain perspective – his own? but we must not pretend to that – would seem to be even more troublesome. He is a creature going out – albeit forced under the pain of corporal punishment (what other words *are* there?) – as a principal agent in the bringing into slavery, if not to their death, other creatures of his own kind.

I will readily admit that we cannot know what a horse thinks. Let's instead place a human in this predicament. Within Australian terms, would it be permissible to remind one's reader of the use of Aboriginal trackers, say, to 'bring to justice' other Aboriginal people wanted for 'crimes' against white persons or their property? Would it be permissible to wonder how these trackers felt about what they were being forced to do? And about how other Aboriginal people might see it? Or should we think instead of a slave – or a worker in a concentration camp? – who advances his/her own concerns, or perhaps simply ensures his/her own survival, by assisting the slave owner in the oppression or slaughter of others? Should we think, say, of collaborators, during the Second World War, working with the enemy, and against their own, to save their own skins?

Doubtlessly some, maintaining steadfastly the species barrier, would see this as grotesque overstatement, even read it as some sort of disrespect to Australia's Indigenous population, to those who have suffered in concentration camps, or to coerced populations anywhere in the long history of slavery, but I believe that that would be a very wrong reading indeed, and possible only to or by those who will not decentre the human from its position of absolute – and lethal – privilege, and therefore still see non-human animals as of an order apart, and below.

But perhaps – and setting aside the point, which of course 'crashes' a number of those I have just made, that the horse concerned was not at any point given any such choice – the idea has already come across. 'The Man from Snowy River' may be an (some would say *the*) iconic Australian poem but it has, as do many iconic Australian things, its dark side. One might even – if to do so weren't at the same time to return it to the overweening solipsism of the human – go just the one step further and suggest that, at its core, this poem is a kind of re-enactment (utterly unconscious, as some of the most eloquent re-enactments are) of something in our history which it might be best to leave the reader to figure out for themselves.

I have foreshadowed, in my title, some discussion of 'cracks' in this poem. Elsewhere (2007) I've written of the 'Warp'. The two – cracks and the Warp – are not unrelated. Indeed, I think it is clear that the 'cracks' I see in this poem – since they are cracks only, not yawning fissures – are a version of the Warp. The Warp, as I have presented it, is a strain between the form of a text and its circumstance, such as we find when a northern hemisphere, Anglo-Celtic or Euro-centric literary form attempts to deal with very different (e.g. Australian) landscape and circumstance: a postcolonial phenomenon, if you will. But here, while the cracks are doubtless in some part this, they are just as doubtlessly in some part something else. I would like to say that what we witness is some deeper, unconscious consciousness – if I may be permitted the oxymoron – of the cruelty and inhumanity entailed in this kind of crime against kind, but, things being as they are, I can't plausibly claim it as anything more than some combination of evidence of an absence or failure of thought – the linguistic invisibility of non-human animals – and just bad writing.

The cracks begin with the 'cracks' themselves, though they are perhaps the least of it. Their triple- or quadruple-ness, as attested by the *Oxford English Dictionary*: 'cracks' as in 'crack horsemen', most obviously, and also, sympathetically, as in cracks of the whip, but also as in boasters, people who *talk themselves up*, and then as in fissures in the skins and schemes of things. But this is just word-play, would mean little or nothing unless it were part of a pattern. We have seen stronger cracks, in any case, in the initial, self-contradictory description of the stripling's horse, and later in the tension between his fiery-hot courage and the description of the brumbies after his chase. And there is that later problem of positioning (that the stripling, at the height of his ride, is both *with* and *following* the horses he is trying to capture) in 'he was right *among* them still, / As he raced across the clearing *in pursuit*', and the sudden and momentary shift of tense, for the rhyme's sake, in the next two lines, which ironically allows the poem to glimpse its own legend:

> Then they lost him for a moment, where two mountain gullies met
>     In the ranges, but a final glimpse reveals
> On a dim and distant hillside the wild horses racing yet,
>     With the man from Snowy River at their heels.

There is also something in the description of the stripling himself – a touch of confusion, as he is introduced (a matter of the stripped, telegraphic grammar often necessary to fit content into form), as to whether the second line refers to the horse or to the youth:

> And one was there, a stripling on a small and weedy beast;
>     He was something like a racehorse undersized.

Technically, the referent of 'He' should be the boy. By the time we get to 'racehorse' the confusion is arguably gone, although just as arguably 'the small and weedy beast' is now ghosted with the boy and vice versa. Paterson could have avoided this confusion by using 'That' instead of 'He'. Whether he avoids this solution deliberately, to *ensure* the confusion, can't be known, although, just two lines later, the last four lines of this verse paragraph, presented as an independent semantic unit, seem to bear the same subtle ambivalence (i.e. could refer as readily to the stripling as to his horse). I've quoted them once already but a second time won't hurt:

He was hard and tough and wiry – just the sort that won't say die –
  There was courage in his quick impatient tread;
And he bore the badge of gameness in his bright and fiery eye,
  And the proud and lofty carriage of his head.

There is a likelihood that this is deliberate, of course, since at so many points in the poem – this one central among them – the stripling and his horse are doubled, seem almost to osmose through the species barrier, and since it is not just the stripling and his horse that are subject to such slippage. What happened to the colt? He *begins* the poem – his escape/disappearance is the cause of its coming into being, and his value would surely make him the principal object of the pursuit (indeed the poem tells us that quite directly) – but then, quite literally, he disappears once more, is never again mentioned after this initial appearance (if we can call an absence an appearance). Instead there is a kind of elision. Attention shifts to those other colts – the stripling and his pony – as if they, the latter in particular, become his avatars, pursuing him and *being* him at the same time, emphasising all the more the matter and problem of *kind* that haunts this poem. As if the poet were somehow nosing at the species barrier or, more likely, the species barrier nosing at the poet.

Elsewhere it is a matter of particular words themselves. (Can words, individual words, have cracks? But we saw it in the word 'crack' itself, that *shaling* of the signified, its splitting into its various layers.) Take 'weedy', for example, or that interesting oscillation between – better, since it has tentacles that reach far into the poem, call it a complex in part comprised by – 'bloodhound' and 'cur'.

From a certain angle, to be described as 'weedy', as the stripling's pony is described, may not be quite the insult or criticism it might be intended to be. Weeds can be tenacious, very hard to get rid of, and the actual problem they represent has far less to do with any inherent weakness than the manner in which they contaminate the cultured space of the garden, in which they are, as it were, some combination of unwanted immigrant and threat to the breed-stock, if only in the sense of competition for resources. In effect it is not their weakness but their strength that is the problem, a strength that is hard to separate from their wildness. But of course we don't see this, galloping as we do through the poem, and the use of 'weedy' to designate weakness – thinness, fragility, underfedness – is after all quite established.

And then 'cur', in the penultimate stanza of the poem. It's only recently, in language-time, that the term has become so unequivocally derogatory. Earlier in its history it's been simply a term for a dog, and in some places more specifically for a shepherd's or a goatherd's dog. There is, too, in its more contemporary form, a pronounced connotation of the mongrel, albeit strongly in the negative, as if the mixture of bloodlines were somehow a weakening or contamination – something which, as it happens, this poem is very much in two minds about. The stripling's horse has 'a touch of Timor pony' but is 'three parts thoroughbred at least', 'And such as are by mountain horsemen prized': good to be pure-bred ('three parts thoroughbred' is its strongest feature?); good to be a mixture (the 'touch of Timor pony' is good); great if the mixture has a lot of *purity* in it.

But I was speaking of 'cur'. The mountain horse has just been defined as hybrid. To say that 'never yet was mountain horse a cur' is therefore contradictory in this second sense (we have seen, a good way above, that it is contradictory in its primary, ostended sense). And surely the word 'bloodhound', just a few lines before it, means that it is contradictory in a third, the sense of 'cur' as *dog*. Obviously a horse is not a dog – we have to extend some poetic latitude here – and in fact in this sense Paterson has merely stated the obvious. But in another sense he has just contaminated it. (One of the problems of poetry: grab at any passing metaphor and you might find you've invited in more than you can handle. 'Bloodhound', for example, is problematic in another sense also, in that, while it may bring in the relentlessness – the *doggedness* – of pursuit, it also brings in a matter of scent, of trace, of *tracking* that is, on the one hand, quite inappropriate to the particular action here [where the stripling never seems to lose sight of – is frequently in amongst – his quarry], and on the other comes close to pointing up one of the guiltier secrets, the represseds, of the poem.)

There are others I might have chosen, but perhaps these three words are sufficient to establish the pattern, by which word after word in the poem has a momentary force that destabilises when, reading *against* the driving rhythm and the deceptive pleasures of the rhyme, we examine it further. Of course it could be argued – I argue it to myself – that a great many texts can be read in this resistant way, and even that it is in the nature of close reading in the first place to expose the kinds of contradictions and inherent tensions that this reading has exposed. The patent anxiety over bloodlines, purity, contamination and *authenticity* in the cracks just examined, however, seems of a particular kind – we might call it physio-ontological – and would seem to articulate, deepen and extend the kind of contradictions and tensions we've found more crudely and far more obviously in Paterson's poetry elsewhere. Their extensions into and re-enactments of invader relations with Indigenous Australia are, I think, fairly obvious. That these occur so extensively through the vehicle of a non-human animal suggests that the species barrier, albeit almost utterly unconsciously, is just as much a matter of unadmitted concern.

We are animals. We are language. Human abuse of non-human animals, universal and almost universally accepted as it might be, is always – since it entails so much denial, so much repression of *kind* within the self, so much deep *contradiction* – so consistently accompanied by psychic distress that we might argue this distress as a constant of the human mind and even a part of the definition of the human. When a text turns consciously towards non-human animals – turns, as does 'The Man from Snowy River', and tries to hunt one down – it can't much surprise us that this psychic pressure begins to manifest linguistically. And if, as anticipated above, it *is* objected that all texts have cracks in them, and that all texts can be treated like this, then what does that say?

# Animal Dreams

*the phantasms of sleep do commonly walk in the great road of
natural and animal dreams*
  – Thomas Browne, 'On Dreams' (c.1650)

WE CAN'T CONTROL OUR DREAMS. Most of the time we can't even
remember them, although some tricks can help. Write dreams down as
soon as possible, or tell them – even if only to yourself in the morning
shower – and you'll remember them better: translating them into words
in any form will lodge them more firmly in the mind. But these are
only the dreams we wake from. How many do we miss because nothing
wakes us while we're in the midst of them? Do they lodge, somewhere
in the brain, a deep memory we can't consciously access? Do they
influence us, in ways we will never know?

Doubtless, too, the dreams we remember are incomplete, *because*
they're the dreams we wake from. Where were they about to take us?
What would have happened next? Are there *complete* dreams? What
would *they* be like? How could we ever tell?

An experienced observer, watching us night after night, able to
determine (from our Rapid Eye Movement?) when we are dreaming,
waking us to ask us, getting us to turn our dreams into words, might
help us understand better the patterns of our dreams, but then
(Heisenberg) the act of observation might alter the thing observed.

Start interrupting dreams and soon they'll become dreams about interruption. One thing we do know about dreams, after all, is that they are hypersensitive to stimuli surrounding the dreamer. A door slams, or someone drops a book in the hallway, and a dream is likely to end with a gunshot, a stumble, a fall.

No. We can't control our dreams. This is their mystery and their power. For some, they're simply the mind sorting out the various stimuli of the day, tidying itself, readying itself for the next, not communicating anything in particular, but for many others they seem to be *communicating*. Psychoanalysis has always seen them as a mode of access to the subconscious. Although it says little about non-human animals,[1] Freud's longest book is *The Interpretation of Dreams* (1900).[2] As if dreams present their material with a kind of honesty of which the dissimulating waking self isn't capable. And throughout recorded human history, from civilisation to civilisation, dreams have been taken as signs, auguries, indications of what is to come, or guides to the meaning of what is already here. Most of us, for example, think that, whatever else dreams are, they are indicators of our deeper wellbeing or unease.

And of course there is the relation of dreams to language, and to narrative, in the first place. Lacan tells us quite persuasively that the unconscious is structured like a language. In remembering a dream, we turn it into narrative. But how much about it do we push away by doing so? Perhaps, in constructing the dream narrative, there is as

---

1    A term which, although in substantial agreement with him, I use *pace* Derrida: 'I avoid speaking generally about animals. For me, there are not "animals". When one says "animals" one has already started … to enclose the animal into a cage. … [T]o say "animal", and put them all into one category … is a very violent gesture' (from a filmed interview, at https://bit.ly/ 3htTr4m). The point is made more extensively in *The Animal That Therefore I Am* (lectures delivered in 1997, first published by Fordham University Press, 2008), 29–34 in particular.

2    For Freud, as far as I can tell, non-human animals in dreams are predictably and anthropocentrically symbolic and have little or anything to do with a return of or to the body; for Jung, on the other hand – no less anthropocentric but a little more receptive to such a return – animals can be assistants in the hero's journey, through the dark night, towards completer and more integrated being.

much information – if dreams *are* communication – in the details we pass over as there is in the details we choose. We may remember that in our dream we passed through a door into a long, narrow room, but might it have been the door itself that was 'talking' to us, the grain of its panelling, the shape of its handle? (One idea that has always made sense to me is that, whatever else the different parts of a dream might represent, they are always also in some way oneself. The door is oneself, the grain of its panelling is oneself; you are the dog, the dog is you.[3])

Dreams border on another place. The place of our unknown. The repository – the moraine – of the vast processes of relegation and *differing* that are necessary for us to be able to 'see' and 'know' in the first place. Although, from a dream, we may only remember what we 'recognise', the dream-quality, surely, comes from the presence of – the making almost nightly almost conscious – the unknown that enables our knowing, the humus from which we cultivate our waking consciousness. Individually, but also – since nations, cultures, are made up of individuals – as a culture, a nation, even a species. How do the dreams of an individual relate to the dreams of a race, or a culture, a nation? Does anything, other than a random metaphor, connect, say, a 'national' dream with the dreams that a sleeping individual may experience?

Certainly there are some strange and haunting indications that one's dreams are not entirely one's own. Once, in a short story I was writing (1985), I described a dream I had just had. 'He begins slowly,' I wrote, 'at a man's pace through the veldt, the stands of low trees':

> The sinews of his forelimbs, the fur on his shoulders, ripple about him … Gradually his speed increases, he begins to lope. The dust rises gently behind him. He notices less now, his eyes focussed on an invisible, racing point some five or six yards before him, choosing his path as rapidly as he, now running, devours it. Thin legs flash in panic from his field of vision [and] the flight of smaller animals creates a strange wake in the grasses behind him.

---

3    Though what then to make of a dream in which one finds oneself walking down a busy inner-Sydney street beside a dog who, through nasty accident or an interrupted surgery, has half his/her brain exposed?

Now, in smoother, more open country, he moves yet faster, his paws marking the ground more lightly and at wider intervals, that slow rocking beginning within him that marks his highest speed, his body now a wave barely grazing the ground beneath him. Suddenly there is a wheel, an arc in the air, a panting, a lusting as he tears, the spat of blood in his nostrils, the rip of skin and hair. Ripples, large at first, spread out in the plain, disappear at last, into the far trees.

*I was there: there were two of us. The prey we caught was our own sorrow. Our loneliness. What a feast it was!* (24)

What a surprise, then, to find in an Australian novel, published a couple of years earlier, but which I had neither read nor read about before, passages such as the following:

The animals were there. They'd been waiting all night to reveal themselves and they sprang out now, so clearly that she could see individual hairs on their coats and the blank ferocity in their eyes. Then a siren blared. But the beasts, not alarmed by it, continued to tear at each other. Judith was gasping. She forced her eyes open and knew then that the bedroom-extension telephone was ringing

and:

The beasts were waiting, wrestling with each other in the long, bleached grass. Then they set out to hunt, padding through the stretched shadows of forest twilight. The creature they felled was only half-grown; they had just captured it when it vanished. In rage at being cheated they turned away from the empty patch of grass, and the larger tiger – who with dreaming eyes had groomed his mate with his tongue ... – reared up and leapt towards her.[4]

What could I make of this coincidence? Was the same emotion, in each author, producing the same image, the same dream? And how much in these dreams is doing the signifying (beyond the beasts themselves

---

4    Blanche d'Alpuget, *Turtle Beach* (1981), 33, 45.

– the big cats – there is the coincidence of fur, of grasses)? It would seem clear that, for each author, the dream is sexual, but why has the subconscious, for each of them, chosen the big cats as its vehicle? Is this association an ancient one, emerging from the deep mind – Jung's universal subconscious? – or has there been some common external, cultural source, something as simple as a movie, a song, an advertisement?

But all this is preamble. A preamble to a preamble. A great deal has been written on the subject of dreams, and as a taxonomist of dreams I am a novice. I want instead, and as indicated by those just adduced, to look at a particular kind of dream, which I describe as 'animal' dreams. And, having no one else's that I can approach with any confidence, though I am sure we all have versions of them, I will, in the hope that they are not so unfamiliar to others, use my own, as much as reminders of dream function as for any particular significance in themselves.

Once, in a time of great distress, I dreamt of an aviary full of dying birds. I went, in this dream, into the cage, trying to save them, but every bird I took into my hands died as I held it. And once, later, in a time of great happiness, I dreamt of a similar cage – perhaps it was the same one – and whatever bird I took into my hands revived and flew free. I interpreted these dreams, in accordance with my understandings at the time, as dreams of impotence and renewed potency, not necessarily sexual, although that is the twist Freud might have placed on them. The death of a loved one, the end of a relationship, the loss of a job: these things are just as likely to stimulate feelings of impotence. And the impotence – the interpretation of impotence – that I placed upon these dreams was, perhaps needless to say, a self-centred and anthropocentric one.

Now, I think, my interpretation would be slightly different. I don't say that it would not be still to some extent self-centred and anthropocentric. These things are so hard to avoid, I think, that the very claim that they have been avoided is tantamount to evidence that they have not. But the mirror, let's say (let's hope!), has now been slightly tilted, to catch at least a glimpse of something else.

The two dreams I next want to talk about occurred on successive nights many years later. The situation they address is a medical one that had seen me taken to hospital and, for twenty-four hours or more, had seemed to threaten my life. The first occurred on the night after this crisis passed. I transcribe directly from my notebook:

A large wooden door in the white, stone-paved street of a sunny Mediterranean town. The door gives on to a long, narrow room, dimly lit from a window at its far end. I enter carefully. There's a small creature in there that will be trying to escape. Pup- or kitten-like, but more solid, lower to the ground. A small Tasmanian Devil perhaps, but with softer fur. I enter and leave the room a couple of times. On one occasion the creature struggles to get out past my legs but I manage to keep it in. On another I see it clinging, flattened, to the inside of the door. I find a couple of young people in the room, at a table at the far end, and warn them that they must not let this creature escape. On a later occasion, however, it gets out and flees ahead of me down the street, racing off around a corner to where there are outdoor cafes. I manage to follow for a short while, lose sight of it, and double back to find it only about twenty metres from the wooden door. I pick it up by the scruff of the neck. It struggles, trying to get its head around to bite me, but I manage to stay clear of its teeth and claws. I hurry off and take it back to the narrow room. There I find a red plastic box the size of a three- or four-litre food container, with a white lid. The box has slits in its sides, as if it were a kind of cage. I place the animal in it and take it back outside, to return it to whomever I have been looking after it for, for clearly it's been in my care. I feel – and hear – a sudden scrabbling from inside the box. It ceases. A few seconds later I open it. Inside there is nothing but a couple of fresh-gnawed bones. As if the animal has eaten itself, or been eaten by something within the box, invisible to me.

When I think of what was going on at the time – something within the box of myself, invisible to me, threatening to devour – this gist seems clear enough. But this crisis might not be the only factor. My daughter had earlier asked me to look after her chihuahua while she went interstate. And I had, a month before, been asked to look after a friend's small poodle for an afternoon. The poodle, who had never been left with a stranger before, had escaped around my legs as I opened the front door to accept a delivery, and had run off down the street, directly towards a six-lane highway. My dream, remembering the poodle's escape, may have expressed a fear that I might not be able to look after my daughter's dog. Perhaps, putting the two readings together, my own physical instability, unsettling me, created parallel fears. I might have settled for this, had it not been for the dream on the night following.

'I don't remember', my notebook reads, 'any details of the dream before this scene that might have explained how I came to it':

> but I am in a building, leaving an apartment on the top floor. I've just come out, with a dog I have to get downstairs somehow. I've left my keys inside: locked myself out. And there are no stairs, only a hole in the floor giving on to a set of levels more ladder-than step-like, wide enough for only one of us – me or the dog – to pass through and climb down. I lower him through, past the first step, but then realise I've created an impossible situation. I am responsible for him, but should have found some other way to get him down. Now, as it is, I can't get him back up, around the first level, and if I let him go there is (a) no certainty that he'll drop only to the second step, and not fall all the way, and (b) no way that, if he *did* land safely on the second level, he could then get down to the third. I've trapped us both. His life seems to be dependent upon me, and my grip is getting weaker.

The hole in the floor here very likely represents a link to a deeper level of the self. And I have pushed down – trapped – the dog there. Once more, of course, there are other explanations (it could be the escaping poodle again), but what intrigues me is the presence of the animal in the first place. If we have to allow the poodle explanation some space, the question becomes *why* my subconscious chose *that* for its subject.

Far from trying to argue, as so many have done, that it is the emotions that *separate* us from non-human animals, in these dreams it seems that in some way the emotional complex I was going through, in a period of personal trauma (it's not for nothing that the word for dream in Freud's language [*traum*] and *trauma* are connected) was stirring up the animal, whether as *the animal in myself* that I was apparently unable to protect, or, as I'm more inclined to suspect – and this *is* different – the voice or being of the animal *of* my self, surfacing when emotional disturbance rendered the consciousness more porous.

If we were to attend to our dreams more closely, and more particularly to non-human animals ('the animal') in those dreams, mightn't we find that, rather than separate us from other animals, the emotions, when they become sufficiently strong – when they charge us like frightened elephants – are exposing and even leading us *to* the animal in ourselves, as to some inescapable and irreducible core?

Is that my point? Well, yes and no; mainly no. I could try to analyse this kind of material further – my musings so far are only a beginning – but, as I've said already, I am not a dream psychologist and don't wish to present myself as one. As far as the non-human animals we live amongst are concerned, there are more urgent needs to consider. It may be best to regard the foregoing as a kind of preliminary massage. While it may not be true to say that renewed emphases upon and re-visionings of the animal within us can never go astray – they can go horribly astray – some gentle manipulation of the psychological tissue that holds (and restrains) that core may help to stimulate our sense of wider kinds and operations of the unconscious in the society around us, and a sense that, *as collectives*, we are not always as aware of why we do and think things as we think we are. Some of our public policies, for example, and our reactions to them, may not be quite as straightforward as they've appeared to be.

Nations, it can be argued – much as Wilhelm Reich does, in *The Mass Psychology of Fascism* – are constructed a little like the individuals of which they are comprised. That is, just as, to perform as individuals on the social stage – to come to have the *identity* that can enable us to do so – there are parts of our wider selves that must be repressed (and that therefore whelm in our dreams), so too, in assuming and maintaining *its* identity, *its* public face, there are parts of the nation that must be repressed, even abjected. And it is arguable that the repressed whelms in a nation's dreams – in its advertising, for example, and its popular culture (the *metaphysics* they seem to provide), for these are some of the most evident forms of a nation's dreaming, but also in some more obscure ways that, like the personal dreams I've discussed, may hint at darker texts.

Early in 2011 the Australian nation rose in indignation at the treatment of Australian cattle in Javanese slaughterhouses. In footage broadcast by the ABC on 30 May of that year we saw creatures being slaughtered without prior stunning, throats being cut with hideous inefficiency, other animals having to stand by, watching this being done, knowing that it was about to happen to them.

What else may have been happening was not so clear. Albeit cynically and only until our compassion fatigue had set in, the government suspended live cattle export to Indonesia. Some of those who rose in protest, confronted by the cruelty upon which its production is based, went so far as to give up eating meat (for days, even weeks ...[5]). For most, however, the issue seemed largely a matter of method. The strange oxymoron *humane slaughter* circulated widely, as if there could ever be such a thing.

The psychological hydraulics – the shuntings between different parts and levels of the national mind – involved in this process were intriguing. Abjection was at work – that process by which identity is maintained by the expulsion of what it cannot bear to face. We may have called for, and achieved for a time, an end to live export to Indonesia, but somewhere along the line the cattle had been replaced or underlain by something in our subconscious.

In our anger at the treatment of the *cattle* we'd exported, we were actually exporting our *shame* – processing it, as we had been processing the cattle, off-shore – and xenophobia and racism were a part of it (' off-shore processing': another fragment of this strange poetry). We were criticising barbarity, yes, but we were also criticising Javanese, also criticising Islam. And what we were criticising was bolstering what we think we are. And yet, of course, these very things we were criticising are not only amongst us but are part of what we are, and I do not mean only Islam.

In the years that have passed since the first heat of our objection died, numerous abattoirs have been shut down (albeit only temporarily) in New South Wales alone, for acts not much less barbaric than those we witnessed on our TV screens in 2011. Drive from Sydney across the Blue Mountains and out onto the great western plains – to shift perspective slightly – and you enter, yes, some wonderful country, but

---

5    And some permanently, I do not deny that. At least one now well-known
     farm animal sanctuary is said to have started in its founders' horror at these
     scenes.

you also encounter, placed out of the line of sight, obscured by isolation and distance, penitentiary after penitentiary, 'correctional centre' after 'correctional centre' (within them, at a further level of remove from the national sight, a massively superaverage proportion of Indigenous Australian prisoners), and, of course, animal factory after animal factory, abattoir after abattoir, as if, as well as providing something of the nation's dream, the 'outback' also provides a cloaca, a place to which to relegate what might otherwise become nightmare.

What can we do with this? It may be too simple to say that our horror at what we saw on television that night in 2011 was in some part a horror at what we have done to the animal in ourselves and that – despite all the evidence that it (this thing that we have done, that we do) is *here*, is *us* – we *had* to export it, *had* to flush it from ourselves, in order to maintain our image of ourselves. But in a country where cruelty towards other animals is endemic and even institutionalised (the mass slaughter of kangaroos – our national symbol – annually in the national capital may be itself symbolic, but is only a tiny part of it: in addition to continuing eradication programs of dingos, rabbits, foxes, cane toads, feral cats, feral pigs, say, there have recently been extensive culls of camels, wild horses, donkeys, water buffalo, corellas, Indian mynahs, crocodiles, goats, ibis, koalas, flying foxes ...), we can hardly dismiss the possibility.

WE CAN FIND OURSELVES STUMBLING, as if the rational, ethical creature that we think we are has one foot not properly functioning. When non-human animals are involved, particularly, the wind can blow in contrary directions, as if something deeper than consciousness, deeper than ethics were operating. To many (for example), one of the most alarming aspects of the 2011 footage was that the cattle had not been *stunned* before slaughter, as if rendering animals insensible before cutting their throat were somehow kinder than allowing them to remain conscious, aware of what was being done to them. I don't necessarily dispute this, but it's less straightforward than at first appears.

Does extending a little consideration towards animals before slaughter make that slaughter any the less barbaric? And is it to the *non*-human animal or to ourselves that we are concerned to be most kind? Not having to view the slaughter in its full horror and cruelty also lessens the pain *we* feel, as observers, in those rare instances when we choose or are forced to look upon the price and cruelty of our food choices – helps us to hide from ourselves our complicity in the very practices which repel us, and so to reconcile ourselves to the continuation of those practices.

And was I alone, in this moment of national outrage, in thinking about dignity, sovereignty? Was it outrageous to wonder whether there might be a kind of respect shown to animals, non-human or human, in allowing them to be aware of their own dying? In not depriving them of an awareness of what is being done to them? *In allowing them to witness* our barbarity? (But these – dignity, consciousness, awareness of one's own dying – are things we are led, by a great many philosophers and scientists, and contrary to the evidence of our own eyes, to believe non-human animals are not capable of.[6])

Live export is no isolated conundrum. Early in 2014, occasioned by the death from shark attack of a surfer, there was hot discussion throughout the country concerning whether or not, firstly, to seek out and kill the shark involved, and then whether to 'cull' Great White sharks generally. Hundreds of sharks died from consequences of that one fatal attack and the few non-fatal attacks which, that summer, had preceded it – a detail, however, which must be considered hand in hand with another, that there had, as of January 2014, been only two hundred and seventeen recorded deaths from shark attack in Australian waters since records began in 1791 (compare with the 1193 road deaths in Australia in 2013 alone). Even the family of the surfer involved pleaded against the cull, saying that those who enter the sharks' medium know the risks, and that sharks should not be punished for behaving in a manner that can be entirely predicted – usually through mistaking the surfer for some other large prey. But something else had been stirred and what resulted, as testified by its rank absurdity (the hunt would be more correctly termed a kind of bizarre *reprisal*, directly along the

---

6   See, for example, Martin Heidegger, in 'The Thing' (1950), trans. Albert Hofstadter, in Heidegger, *Poetry, Language, Thought* (Harper and Row, 1971): 'The mortals are human beings. They are called mortals because they can die. To die means to be capable of death as death. Only man dies. The animal perishes. It has death neither ahead of itself nor behind it' (176).

lines of those undertaken by, say, the Gestapo, in killing ten, twenty or more villagers for every German soldier killed by a partisan: as if sharks might be aware of history, or understand the basic principle of eyes-for-an-eye and will adjust their behaviour accordingly), was a matter of mass-psychological momentum.

Here again an ' ethical' matter abridges something else. I'm aware that every time I've swum out into deeper water, away from other swimmers at an ocean beach, I've worried about shark attack; I have been *aware* that I was taking that *risk*. The situation is analogous to that of someone who, aware of a 'vicious' dog in a yard, or 'dangerous' bull in a field, chooses to enter that yard or field anyway. Is it *ethical*, then, knowing this risk and taking it anyway, and finding that one *is* attacked, to seek the death of that dog, that bull? The shark cull couldn't save the life of the person who'd already died, and there was no way of knowing beforehand, that is, without examining stomach contents, whether any shark subsequently killed was the shark that killed that person. The culling took on a rather different aspect, became, in some part, an attempt to annul the child's – and adult's – fear of the ocean and what might come from it, a part of our deeper psychological structure, our deeper dream.

I suspect I'm hardly alone in my experience of deep-water anxiety. I suspect, indeed, that our fear of the ocean – fear of deadly attack from the water – is so deep, so widespread, and has its origins, begins its *imprinting*, so commonly in early childhood, that it is a significant factor in the national psyche. Nor is it a matter of sharks alone. There are also, for example (and setting aside that actual deaths from these creatures have also been fairly minimal), blue-bottle jellyfish, who drift in to beaches in their thousands and inflict painful stings. And, further north (I write from New South Wales) and far more dangerous, there is the Irukandji, a jellyfish so deadly that, on some of our most beautiful beaches, one dare not go into the water for much of the year. And, of course, in many of the same areas, and extending into coastal rivers and fresh water reserves, there are crocodiles, deadly attacks by whom are reported almost as often as shark attacks.[7]

An extensive culling of Great White sharks, all this is to say, might address something in reality, but as large a part of it would seem to stem from a darker and relatively unexamined place in the national psyche, a place the latent force of which extends beyond non-human creatures to national policy, explaining at once the fervour of our coastal fortification against possible attack by the Russian navy in the 1840s and '50s and our vulnerability to – and preparedness to elect governments which will undertake – propagandistic manipulation against those who endeavour to reach Australian shores (as so many of us or our ancestors did) by boat in search of asylum, as if they, too, were some form of deadly threat from the sea.

Go back, then, to our reaction to the revelation of the brutality of an Indonesian abattoir in the slaughter of animals *sent there in our name*. Can we separate this utterly from the question of shark reprisals, or any other element in the soup of our abject? These sendings-away-by-sea, these exportings of shame, these refusals-to-accept or take back, these vastly disproportionate reprisals, these massive repressions of the animal: as far as any one of the phenomena I've discussed might appear to be from another, they can also be seen to be deeply if elusively connected, part of a dark poem we have yet to learn to read. (The people aboard the boats – unseaworthy boats, disposable boats, boats no longer useful for other purposes – packed, as we have been told often enough, 'like cattle' ... )

We *will not accept* our own cruelty, our own guilt. Nor, particularly (one might say) having chosen the diet we have, do we find it easy to admit the suffering animal within ourselves, for to do so would open a dangerous door. These things are not a part of our self-image. And we maintain that image by *abjecting*, by sending or locating abroad, or into our own outback, or, more simply and violently, by killing, that which does not suit or seemingly belies that image; nor will we permit its return, in the form of humans who have been driven out of their *own* home situations by cruelty (physical, political, economic) – consigning them all too often (*once* is too often) to the depths of the sea, as if this were (and it is) some sort of objective correlative of the depths of the psyche. We are pursuing, as a nation – treating the world about us as if it were a projection of the world within, a stage upon which we might

---

7    To date there have been only two recorded deaths from Irukandji stings since records began, and seventy-one from the slightly less venomous box jellyfish. Historically, too, there is less than one death per year from crocodile attack.

enact (as we do) our own trauma – policies that are also in some part dream, some part nightmare. The sooner we unearth, examine, admit and take possession of them the sooner we might begin to limit the damage.

I've written elsewhere[8] of Australia's deeply conflicted relationship with the kangaroo, its national icon, a creature taken up proudly by sporting teams and huge corporations alike, but also massively reviled, mercilessly slaughtered, treated as a pest even by the institutions charged with its protection, scapegoated by a culture whose own avarice and malpractice have denuded and effectually devastated the land and which needs someone, some other being (not sheep, not cattle) to blame (die) for it. And the kangaroo, after all, is an eater of grass (a competitor), and, more galling still, a survivor.

I've written of elsewhere, too, and mentioned above, the attempted eradication of 'exotic' or 'invasive' species of plants and non-human animals that is so key, but also so conflicted, an element in what passes for Australia's conservation policy. In some ways its rationale is indisputable. Feral cats, for example, kill millions of native creatures every day (night) of the year. But 1080, the chosen method of these attempted eradications (Australia's use of 1080, a poison banned in numerous countries, is second – albeit at some distance[9] – only to New Zealand's), has an unpreventable and undeniable by-kill, leads, for all the claims made to the contrary[10] (since a considerable number of Australia's wild creatures are carnivores or omnivores,[11] and many

---

8    See, for example, 'Field's Kangaroo', 'An Exoneration' and 'Cull and Culture', below, 'Roogate' (op. cit.), 'A Roo Battue' (*Southerly* 78.3 [2018], 83–99), or 'Scapegoats', Day 23 of the *100 Days Project*, at https://bit.ly/2FAn7zy.

9    It is widely asserted that New Zealand uses eighty per cent of the world's supply; Australia uses approximately twenty per cent of the rest (four per cent overall).

10    Claims routinely made that 1080 is safe for wildlife are hard to reconcile with (for example) its use to kill hundreds of thousands of Bennett's Wallabies on King and Flinders Islands (etc.). 'Poisoned possum carcasses' – this is Eason et al. (2013) – 'can pose a risk to dogs even up to 75 days after the control operation'.

11    … quolls, Tasmanian devils, bush-rats, goannas, magpies, ravens, currawongs, owls, butcher birds, kookaburras, eagles, hawks, black snakes, brown snakes, pythons, bandicoots, dunnarts, antechinus …

of them scavengers), to the death of many of the very animals it is intended to save.

With the exception of the 3.3 per cent of us who identify as Indigenous, moreover, we Australians are ourselves exotic in this country. Beyond any possible question, the greatest threat to our wildlife is and, since the first days of our invasion, has been, ourselves. Our broad-scale attempts to eliminate exotic animals and plants – plants and animals *we* have introduced – for the threats they pose to native wildlife, are deeply compromised, deeply hypocritical. Asking, as we do (but there is no asking ...), rats, rabbits, foxes, etc., to bear away a burden of guilt we refuse or are simply unable to bear ourselves, we are once again abjecting, once again scapegoating, once again dragging non-human animals into, forcing them to pay the price of, our own psychodrama.

To be fair – Australia, after all, is the Land of the Fair Go – most countries might be like this, faced with similar circumstances. It's an aspect of the manner in which tribalisms – and the larger tribalisms we call countries, nations – maintain themselves and create their mysterious cohesions. But does this excuse any of us? Hardly. It may be too much to say that what we repress outside is an index of what we repress within, and that the animal within us is stirring, confronting – locked in a tortuous, nightmarish grip with – the animal we slaughter, but can we dismiss the possibility?[12]

---

12 I should say in coda that the poodle was safely recaptured, and that the chihuahua enjoyed her stay. As for the dog in the stairwell, I'm not so sure.

# The Fallacies

## Theory, Saturation Capitalism, and Non-human Animals[1]

I WOULD LIKE TO BEGIN BY SUGGESTING that in every poem, as in almost every human artefact, there is, inevitably and inescapably, the trace of slaughter. But that would be confrontational, and I don't in any case have the means to prove it. Especially given that I'm not speaking of the slaughter of humans, though that too is certainly there.

If, however, all poems *were* to bear the trace of slaughter, then some readers might think me about to suggest we should reach a terminal state of our poetic effacement (a curious statement, perhaps, but I'll explain it soon enough) and stop writing poems altogether. In fact I am of the opposite view. The time for effacement is passing, if it was ever here. I'm about to argue for something more like reversal.

I *am* on the side of animals – or, rather, of *non-human* animals, since we humans are ourselves animal, and with the rider that 'animals' is itself an umbrella term, a conceptual violence,[2] itself one of our means of shielding ourselves *from* what we like to think of as '*them*'. And they – non-human animals – are and will be a trace in all that follows.

---

1    The first part of this essay was prepared for a conference on 'Poetry and the Trace', sponsored by Monash University at the State Library of Victoria in July 2008. The second was added four years later, prior to publication in *Southerly* 73.2 (2013).

2    See footnote 1 in the chapter 'Animal Dreams'.

I believe that our arrogating to ourselves dominion over other animals, our having given ourselves the right to exploit and to kill them at our will, is our deep wound and the source of much of our ontological distress, since this dominion has required first and foremost a separation from or deep division within ourselves, whereby we deny our own animality – an animality which, *undenied*, would mean that most of us engage daily in what are in effect acts, if not of cannibalism, then at least an eating of kind – and pretend instead that we are something else, at best that we are animal *plus* something. Sometimes I think that we have had to invent metaphysics – no wonder they tremble so much! – in order to locate and attempt to stabilise this Plus. In the second part of this essay I will speak of *saturation capitalism*. I see the aforementioned deep division within ourselves, and saturation capitalism (and the consumerism which saturation capitalism drives and upon which saturation capitalism depends) as deeply interrelated.

In a forum at which I offered an early version of this essay we were talking about poetics. One might think that poetics are a long way away from such concerns, but that is an illusion. Poetry may not make many things happen, but *poetics* are a different matter. To the extent that the theorists and philosophers have been right in their assertion that we are creatures of language, and that it is through language that we receive and interact with our world, then *poetics*, the laws, customs and styles by which we put one word beside another, are in fact the laws, customs and styles by which we make our world. They should therefore be one of our primary areas of scrutiny.

In the quite recent and rapid development of literary studies – even now they are barely one hundred years old – this sequestration of the human animal and defence of the territory of the Plus has come at once to be symbolised and propagated by numerous forced and rather specious assertions of the autonomy of the text, gathered about and guarded by four towers – towers which in their own turn are symbolised by what have come to be called the Fallacies.

That I assert that there are *four* towers, *four* fallacies, may surprise. We are very probably familiar with three – the *Pathetic Fallacy*, articulated by John Ruskin in his book *Modern Painters* in 1856, and the *Intentional* and the *Affective Fallacy*, the two added almost a century later by William Wimsatt and Monroe Beardsley in their famous essays of those names in the late 1940s.[3] The fourth fallacy – the fourth tower – is arguably of more recent construction, although also the embodiment of a venerable perception. It is what I call the *Representational Fallacy*: that which, on the grounds that language does not give us a hold on reality, that it walls us in our own tower, makes a fallacy of any attempt to grasp the actual with the word. There is not, as far as I am aware, any one essay entitled 'The Representational Fallacy', but most readers will have read essays that approximate it. In some ways it's the most imposing tower of them all.

What *are* these fallacies, and what, precisely, do I have against them? Why do I regard them as in some way themselves dangerous and fallacious?

By the *Pathetic Fallacy* Ruskin referred to any 'description of inanimate natural objects that ascribes to them human capabilities, sensations, and emotions'. To his credit he *does* say '*in*animate', but, as with the later fallacies, this fallacy has been popularised to encompass the non-human animate also, presenting it as philosophically indefensible to extend, to creatures other than the human, human-like sensations and emotions. The natural sciences, for example, are currently torn, as they have long been, over the matter of *anthropomorphism*, and such questions as whether we can speak of non-human animals as having emotions that we can employ 'human' terms for, such as 'anger', 'pride', 'happiness', 'love', 'desire', 'shame' or 'grief'.

By the *Intentional Fallacy*, Wimsatt and Beardsley meant the assumption that the author's intended meaning is significant to the interpretation of a literary work. In the New Criticism's concern to establish and defend the autonomy of the text, their proscription of this assumption came to mean the relegation of *any* concern for the author's intention in the analysis of a work. This fallacy, like the others, has had its extensions and permutations. People now speak of the 'Biographical'

---

3    '[T]he design or intention of the author is neither available nor desirable as a standard for judging the success of a work of literary art', 'The Intentional Fallacy', *Sewanee Review* 54 (1946), 468–88, revised and republished in *The Verbal Ikon: Studies in the Meaning of Poetry* (University of Kentucky Press, 1954). 'The Affective Fallacy', also collected in *The Verbal Ikon*, first appeared in *Sewanee Review* 59 (1949), 31–55.

fallacy, as if there were, as indeed there is, a proscription against using details of an author's life in order to better understand their writing, myriads of author biographies and the entire vogue for author-centred literary festivals, etc., notwithstanding. Revolutionary as it was held to be, Roland Barthes' 'The Death of the Author' (1968) – its ironies never sufficiently recognised – merely re-packaged and re-empowered this fallacy.

By the *Affective Fallacy* Wimsatt and Beardsley referred to the purported error of judging a work by its emotional impact upon the reader, an argument which has considerable consequences for the reception and evaluation of the lyric, and for the functions of *consolation* and *recognition* for which poetry has so long been most treasured.

By the *Representational Fallacy* I refer (to reiterate) to the supposed misconception that there is some inherent connection between words and the things they refer to and, by extension, language and 'reality'. The belief that this is a *mis*conception has been spurred by the philosophy of Nietzsche, elaborated by the theories of de Saussure, and consolidated by Derrida and many others. Ironically, I do not intend to question an essential arbitrariness in the connection between a word and the thing it is taken to represent – that would be absurd – but only to suggest that the famed distinction of *signified* and *signifier* has never quite managed to eclipse, in the *users* of signs, the immense weight and utility of experience and habit. *Theoretically* we may be irrevocably separated, by language, from the world, but this doesn't stop us from ordering a cup of coffee, catching a bus, or from communicating in innumerable other ways quite effectively as we do so. It doesn't stop us, for example, giving a conference paper, or writing an essay, or taking issue with one.

Ruskin described the Pathetic Fallacy one hundred and fifty years ago. Wimsatt and Beardsley described the Intentional and Affective Fallacies over seventy years ago. It might be thought things have moved on considerably from then. But I am not so sure. These 'fallacies' were *identified* by these writers, *named* by them, but they were not *invented* by them. The proscriptions they represent are in fact guard towers of an ancient and perpetually defended citadel. Each tower has been much renovated, and might now go under a different name, but each is still very much in use. Indeed, it might be argued that they are defended now more thoroughly, pervasively and subtly than ever.

A recent permutation of the Pathetic Fallacy, for example, is the warning against and attempted proscription of using animate creatures as metaphors in the articulation of human feelings and situations: we *appropriate* these creatures, so this line of thought goes, for our own purposes, and so fail to 'see' them as they 'are', or, lest this be seen to clash too obviously with the *Representational* Fallacy, at least fail to let them be whatever unknowable thing or being that, based upon our own experience of what we have come to understand as our own being, we might surmise they might or must be. And if we turn to the Intentional Fallacy we have to concede on the one side that the proscription of the consideration of intention was already there, two decades earlier, in the famed 'impersonality' of T.S. Eliot, and in *Practical Criticism* (1926), I.A. Richards' landmark anatomy of the errors of reading, and on the other that it received substantial rejuvenation, but no structural damage, in Barthes' aforementioned essay, again in Derrida's critique of presence, and again, more recently, in a proscription, not so much by critics as by poets themselves, against the use of the first person pronoun.

So, the Fallacies then, and some permutations. What *is* it that I have against them? In one sense I concede that, within appropriate parameters, they are logical enough, as in they conform to a deeper Logos. In as much as one accepts the strategic use of cauterising the text, of asserting its autonomy – and I do think that this should be a stage in or aspect of our consideration of a text – the cautions they represent against certain distortions of reading, by which I mean distractions from ascertaining the basic 'contents' of a text, make perfect sense. I'm hardly alone in thinking that there is less and less attention paid these days to ascertaining the contents of a text, and that any attempt to draw attention to them should be applauded. But once these have been ascertained – it's a bit like preparing a plane for take-off – the others (the referent, the affect, the communication [the poet, the reader …]) should be allowed on board, like the passengers for whom, after all, the plane is in service in the first place. Had Ruskin, Wimsatt and Beardsley used another term – the Pathetic *propensity*, let's say (the Intentional *propensity*; the Affective *propensity*) – I might find myself less critical; but 'fallacy' is a hard term, a barricade, a tower. Theory often attempts to totalise in this way, and often thereby shoots itself in the foot.

On the simplest and most obvious level, the problem is that most writers, 'literary' and otherwise, write in order to communicate. Most know, or have a fair idea of, what it is that they are trying to say, and to a large extent this is *why* they write. It seems paradoxical to them – to us (... to me, as the author of this chapter) – if not perversely obtuse, to have our intentions in this regard sidelined and derided. As a writer, a poet and novelist, I am absolutely concerned for the *affect* my writing will stimulate. I do not write in order to have no affect, and I don't imagine many others do either. The most cerebral writer, the most disengaged aesthete, seeks what has to be regarded as an appropriate and commensurate *affect*, even if that affect is a kind of serene and cerebral detachment. Along with the understanding or at least reception of any *communication* they seek to make, the presence or absence of this affect is the mark of the success or failure of their writing. To have it relegated seems patently absurd.

But these are only first-level difficulties. There are others, of a different order. With the relegation of the writer's life and intention (not, and this is partly my point, that these always gesture in the same direction), the theory of the Intentional Fallacy in effect relegates a very large part of the social and psychological machinery of the text's production. And upon this relegation we face nothing short of the suppression, the effacement, the removal from our view (in many cases the *convenient* removal from our view), of the complex relations of art and life that may well be amongst the most important things that art, and the contemplation of art, is able to teach us: just *what* it is that we sublimate, for example, and why we sublimate it, the prices that are paid for this sublimation, and the prices that, it may be, this sublimation is an attempt to *avoid* paying; the myths that it enables us to maintain, and the damage and the horror that it papers over. But these, close as they are to the very core of writing, are subjects for a different paper.

Before it broadened to take on the Fallacies more generally, *this* paper was called 'The Pathetic Fallacy Fallacy'. Why? We are cautioned not to extend – a better term may be *export* – our feelings to the things and creatures around us, because this is to colonise them, to appropriate them for our own purposes, and so to relegate, deny or efface their distinction, their uniqueness, their essence. Use a bird, say, for a symbol of or metaphor for a human problem or concept (migration, spiritual yearning), and we are not seeing – you are not offering – the bird for herself.

Fair enough. In many cases it would be quite true to say that we are not writing about the bird at all. To reject the use of the bird as symbol or metaphor per se, however, is hardly the solution. Setting aside what is for me the most obvious response – that there is nothing *but* metaphor, that a word in itself *is* a metaphor, that language is, in its very conception, metaphor – some of the uses of the animate and inanimate, non-human world will not be for the defence but for the expansion, not for the support but for the critique of the human; not to bring the creature to the human, but to bring the human to the creature (metaphors – at the risk of appropriating metaphor itself – bleed, so that it is very hard to ensure that the traffic is ever entirely one way).

But that point – simple and yet substantial enough – is only en route to another and more fundamental one. The extension of our feelings to the things and creatures around us is the basis of empathy, and the only kind of empathy we can feel, since the actual nature of the 'feelings' – that word in itself is a metaphor – of these things and creatures cannot be known to us. And this empathy – one is tempted to say 'particularly at this point of environmental crisis', but in truth it should be so at any point – is crucial. We must amplify – develop, articulate and amplify – our feelings for and in the world, sensitise ourselves to it by the only means we can, which is to extend our senses *into* it, and guard ourselves against – consider most carefully – any voices which would have us turn our backs on it, albeit on the most logical of theoretical grounds. It's all very well to derive from the apparent (and very contentious[4]) 'fact' that humans can only know human feelings a proscription against attributing 'human' feelings to non-human animals, but we cannot ascribe to them anything else – that is, we cannot ascribe to them those things that we do not know –

---

4    Flying, for example, as it does in the face of substantial evidence, from Charles Darwin (1872) onward, that there are many emotions we share with other animals, and that should properly be regarded as animal, rather than exclusively 'human' emotions in the first place.

and so, in effect, and a very powerful effect, such a proscription is to propose that we do not ascribe to them anything at all. The respect and veneration of the non-human animal that such a proscription appears to embody arguably mask an effective and, to borrow Derrida's term, *violent* isolation and effacement of the very creature we are supposedly respecting.

Wikipedia, that invaluable but sometimes tricky resort, to which I turned for the date of Ruskin's work, not having that work on hand, offers, in its entry on the Pathetic Fallacy, the following illustration:

> When Xerxes was crossing the Hellespont in the midst of the first Greco-Persian War, he built two bridges that were quickly destroyed. Feeling personally offended, his paranoia led him to believe that the river was consciously acting against him as though it were an enemy. As such Herodotus quotes him as saying *'You salt and bitter stream, your master lays his punishment upon you for injuring him, who never injured you. Xerxes will cross you, with or without your permission.'* He subsequently threw chains into the river, gave it three hundred lashes and *'branded it with red-hot irons'*.

Xerxes' behaviour is absurd, but that is not the point. The point, *my* point, is that, as we are coming to realise more and more each day, the world *is* injured by our actions, and *does* do something very like turn on us, if not in enmity then in something that might serve our own as well as its best interests to conceive of as pain.

But this is only one of the reasons for my discomfort with the concept of the Pathetic Fallacy. There are reasons, for example, – I am thinking, at the moment, of the enduring strength and freshness of T'ang Dynasty landscape poetry – to do with the way the landscape, in which we have lived so many thousands of years longer than in any city, has so deeply shaped the psyche that we cannot be entirely sure what we project *into* the landscape is not in effect a kind of counter-projection. There are reasons, too, to do with addressing and attempting to understand what I have been describing, here and elsewhere, as a deep wound or division within ourselves. But first and foremost, for this particular writer, there are the most simple and urgent of 'animal' reasons.

The barriers the Fallacies place around a text are not unlike the barriers we place around the meat we eat, or at least the barriers we place to our consciousness of the sources of that meat, the abuse, the slaughter, the butchery that are masked by the remoteness, anonymity and architecture of the abattoir, the light and order of the supermarket shelf, the plastic wrap and Styrofoam trays that mean we see so much less of the blood on our hands, the sleights of language which turn pig into *pork*, or calf into *veal*, cow into *beef*, deer into *venison*. How much harder it would be to look, without reawakening some old crisis within ourselves, upon a de-beaked and mangy chicken in a battery farm, or anguished sow in an intensive piggery, or these creatures, later, as their heads are severed or the stun-gun put to their temples, if, in flagrant defiance of the Pathetic and Affective Fallacies, we saw their pain as our own.

As to the Representational fallacy, I will make my approach in three parts. First, and although it may seem as if I have turned to my subject by turning my back on it, a glance at the expansion of what I call *saturation capitalism* and the manner in which it has effectually bifurcated some of the key values and institutions of what we might once have called 'the West', but now are almost universal; second, a glance in the direction of the rapid and massive consumption of the world's limited resources in order to fuel the consumerism that is the primary tool or agent of saturation capitalism; and third, taking up and expanding upon the first part, a brief consideration of the way in which, although we might feel we operate, intellectually, with a measure of freedom of thought, that freedom itself is curtailed by the environment created by such saturation to the point where we must scrutinise some of our most cherished contemporary ideas and assumptions for the manner in which, largely unbeknownst to us, they may serve as agents of this saturation. Since, as I have suggested already, the world is put together by words, some of these cherished ideas may be literary ones – or, more specifically, may be in the realm of literary theory over and above the fallacies already discussed.

One by one, over recent decades, the barricades against saturation capitalism have been falling, to the point where this saturation could be seen as the dominant force of our age. Democracy, for example – that pillar of Western thought and liberty – has been brought to or forced upon regime after regime.[5] So much the better for the world as a whole, one might have thought (or been *conditioned* to think). But it is, of course, a poisoned chalice. The first part of this paper was delivered four and a half years after the capture of Saddam Hussein and the commencement of what the West then referred to as the 'economic recovery' of Iraq, a double-sided blade since it has become clear that it referred not only to the return of currency to the hands of Iraqi citizens, but to the West's economic re-colonisation of Iraqi territory. It was delivered less than a year after the West's defiant establishment of the Republic of Kosovo, and almost twenty years after the partition and redistribution of Yugoslav republics as nurseries of Western capital.[6] And as, in August 2012, I sat down to begin this second part – the coincidence, salutary as it may be, was no more than that, coincidental – Hillary Clinton, the then US Secretary of State, was marshalling 'the West' to begin drawing up plans for the 'economic recovery' that she saw as part and parcel of the imminent democratisation of Syria, itself, and however appalling its regime, also a further domino falling to saturation capitalism in a process we've referred to as the Arab Spring.

Democracy, this is to say, for all its possible benefits, has also been the harbinger of saturation capitalism, saturation consumerism. Each new democracy opens a new market, not only offering up a new populace to penetration but, by *atomising* that populace, ensuring the depth and efficacy of the penetration. Advertising – exposure to the world's wares – has ensured that the first desire of the politically re-enfranchised individual has been for economic empowerment, the power to purchase, not realising that any empowerment they are thereby granted is also the power to serve, that is, to subject themselves. So the British East India Company 'liberated' India. So the Opium Wars 'liberated' the hidden kingdom.[7] And attendant to this process is the

---

5   In *The* End of History *and the Last Man* (1992), Francis Fukuyama argued that Western liberal democracy may be the end point of human political/ cultural evolution and, universally, the final form of government.

6   See, for example, the Serbian/Canadian documentary by Boris Malagurski, *The Weight of Chains* (2010).

7   The new empires, as we are taught, bear the (household) names of multinational corporations, albeit that the true centres of power, their imperial centres, may mean nothing to most of the households they enslave.

cynical redeployment of that other sacred pillar of the West, individual liberty, the right and power to become and to express oneself, consumer capitalism being based, as it is, on appeal to and exploitation of this very desire. One *becomes*, it would lead us to believe, through *acquisition* of those things which best reflect our individuality. While it may be too much to say that, one by one, some of the most sacred rights and institutions of the West have become compromised in this manner – becoming hosts to the viruses of globalism, economic imperialism – at least two of them, in the manner just described, have become disturbingly bifurcated, and we should surely examine carefully in this light such others as we are able to identify.

It is just as indisputable that saturation capitalism, and the continual over-consumption which it encourages and upon which it depends – which is its vehicle and agency – requires a massive consumption of the world's resources, massive destruction of rainforest, massive overfishing, massive destruction of habitat, and pollution to the point of significant climate change. It is perhaps a fond point to think that if a household were truly conscious of the resources it consumed and the damage its consumption did to habitat then it might begin the better to resist such saturation, but (a) it is of the nature of saturation – hence my use of the term in the first place – that it either leaves little or no room for such thought, or simply and effectually disempowers such thought by the strength of its appeal to contrary desires, and (b) any decline in market such growing consciousness might achieve is readily compensated by expansion into new market areas: as the 'West' grows more conscious of the damage caused by its consumption, other parts of the globe, for the time being, more than take up the slack.

Such massive damage is perforce aided and abetted by a state of mind. *Matter* cannot be consumed, damaged and exploited in this manner where there is care for and attention to it; or, rather, the consumption, damage and exploitation of matter is facilitated by a widespread *dis*regard for and *in*attention to it. (And let us not forget that at the core of this process, this *logic of inattention* – one could say its model and engine – is our slaughter and consumption of the flesh of others, sentient and conscious like ourselves, whose being we have first reduced to thingness.)

It should not surprise us, therefore, to find saturation capitalism attended by a host of 'sympathetic' (supporting) ideas, or, since the immediate context of this paper is a literary one, to find that some of these ideas are in the area of literary theory – ideas which, like those perpetuated by the aforementioned Pathetic, Intentional and Affective fallacies – serve to discourage our intimate and immediate contact with the actuality and physicality, both animate and inanimate, of the world about us.

There is a curious sense of *inevitability* to literary as to any other theory, a sense that, as we move from one theory to another, or move through the various refinements of a particular theory, we move closer and closer to some sort of truth. A part of this is of the nature of theory itself. Every theory attempts to totalise, and so to transcend its own status as theory. But there are paradoxes to consider here, some of them quite simple and evident.

If there *were* any one theory that was somehow ultimately correct and true, then we would not, surely, have such an array of contending theories to select from in the first place. They remain *theories*, after all, and something of our awareness – albeit that it seems sometimes like a subconscious awareness – of this contingency is reflected in our retention, as a descriptor of our literary thinking, of the word *theory* itself. But the sense of inevitability is nevertheless there, and it pays to consider just how this sense comes about, and how inevitable some of these ideas might actually be.

There are – to phrase this very loosely – a great many thinkers and a great many ideas offered to us by those thinkers. How is it that some ideas are taken up and others not? Is it truly a matter of a kind of Darwinian survival of the fittest, of some ideas being right and some being wrong, some feeble and some strong, or are other factors involved? Is there a cultural agenda, perhaps, that selects the ideas that serve it and rejects those that do not? Is it possible that, although many ideas are offered to us, those that come to seem 'true' to us are those which suit most comfortably our cultural priorities (Darwin himself included)? Is it possible that, since the greatest shaping force in the creation of the minds that do the selecting is, increasingly, saturation capitalism, recognised or not, the ideas to which we have given and are currently giving currency are somehow serving its purposes?

As I have just suggested, a *regime* that would pillage the world would be ill advised to encourage love of and care for that world. Is it any accident that some of the most dominant ideas concerning literature's relation to the world are in fact ideas of *dis*connection, *in*accessibility and *non*-relation? Whose – or what – purposes does it really serve to believe that we are trapped in a prison-house of language? Whose or what purposes does it serve to believe that there is no inherent connection between word and thing? Whose or what purposes does it serve to take the 'I' out of our poems (thus encouraging a *democracy of parts*)? Whose or what purposes does it serve to believe that we cannot know anything of the feelings of non-human animals, and that any attempt *at* such knowledge is an act of appropriation?

Whose or what purposes – to turn to one of the principal tenets of contemporary theory – are served by our broad uptake of the idea of the death of the author, as presented in various forms by Barthes, Foucault, Derrida and others? The death of the author, while a major development and liberation for the reader (a democratisation of meaning), was also, if not in those who generated it, then in the way it was taken up and intellectually institutionalised, very quickly made, as it were, an organ of the state, consumerism and repression of resistance – for authors are, are they not, key sites of resistance, critique and scrutiny? Might it not be that this salutary liberation *from* authority has become, eventually, in its institutionalisation – and if we can see that authority can be ceded to and taken up by an economic function – a subtle organ *of* authority? A similar point could be made concerning the severance from origin, critique of authenticity and appropriation of images that are seen as the key features of postmodernism, but – author, authority, authenticity – it's perhaps there for the reader to make for themselves.

They say that theories fly in and they change us. I am not sure how true this is. Firstly, they are produced by people who are produced by a culture; and secondly, they are accepted *by* that culture. If they take off – are found 'valid' – it is the *culture* that approves them. If they flow against that culture in any radical way, then it is unlikely that they will be seen as successful within it.

This must sound as if I am rejecting, or at least criticising, some of the key ideas of our times – De Saussure's account of the linguistic sign, Benjamin's idea of the appropriation of images, Barthes' account of the intertextuality of authorship, Derrida's critique of presence, etc. – but this is not so. These are powerful and compelling ideas, and my concern is not in any way to discredit them. I am speaking not of these ideas themselves but of the ways they have been taken up, the directions in which they have been developed, the agendas and priorities that – albeit largely in the hands of critics, theorists, practitioners and readers quite unconscious that this is what they have been doing – they have been led to serve.

To put this in yet another way, has the direction in which these ideas have been taken been somehow inherent *in* them, or have they been taken in the directions in which they have according to some other factors? An idea is taken up, surely, because it is in some measure *recognised*, which is to say that it has been *found to fit* sufficiently the priorities of the matrix into which it has been released. There may have been many other brilliant observations of this order made during this period that have not been taken up because they were not, or were only insufficiently, sympathetic to this broader matrix. If the vast majority of those likely to be exposed to and to take up such ideas have been deeply preconditioned by, say, saturation capitalism, then it is unlikely that ideas that are antithetical to it will take broad hold.

If the priorities of this matrix were different, might it not have been possible that some of these very ideas might have been taken in different directions? De Saussure's separation of the linguistic sign into signifier and signified, to take but one example, offers us various possibilities. The direction in which we have chosen to take it has led, in effect, to the conception of a linguistic prison-house, and to, *theoretically*, the total isolation of the human mind. An alternate route – directed, let's say, by priorities of connection rather than division, of care rather than exploitation – might have led us to use words with a greater respect for and understanding of their limited capacity and for the huge task before them, the great burden we must perforce place upon them, and to emphasise that sense in which language itself is stimulated by our need *for* and relationship *with* the world, rather than

– *pace* Freud, Lacan, Derrida, Kristeva (etc.) – as something which emblematises the world's absence.

How is it, to turn to a last example, that a critique of 'Humanism' and the 'Natural' – a cautionary reminder that there is a great deal about what we had assumed to be 'natural' and non-negotiable was in fact system, structure, and so to a certain extent negotiable – led to the supposed *end* of humanism? Why did it not result, instead, in more caution concerning our sense of the natural? Why did not some revised and more cautious version of some residual part of the natural *join*, to *remain in the palette of thought*, those things which it seems came so quickly to replace it?

Doubtless a part of the answer lies in that tendency, already cited, of theory to totalise and so to overreach itself. Doubtless a part lies in a kind of oppositionalism – born of Hegelian dialectic – which, ironically, poststructuralism dislocates even as it continues to practise it. Doubtless, too – or is this only to rephrase these same things? – some explanation lies in our relentless *binarism*.[8] But it is hard, surely, to discount the manner in which, first and foremost, it suits and serves so exactly and so sympathetically the broader and stronger current of disempowerment and desensitisation of which I have been speaking.

But enough. Clearly we need to ask – to *remind* ourselves to ask – not just what our new ideas imply and where they might be taken, but what it is that such implications and directions might in fact be serving.

Empathic identification. Intention. Affect. Representation. My point, ultimately, is a very simple one. If our agendas, our priorities were different – if we wanted to promote care for our environment, compassion for species other than our own – *would these things be seen as false?*

---

8    How is it that our sense of the wide-ranging critique that Theory has offered us has become so binary? That we accept, for example, in our thinking about language, if so we can refer to what may in fact be a shutting down of thinking, an opposition of 'transparency' with 'opacity', discounting such other possibilities as, say, Jakobsen's six functions, or the simple observation, known to anyone who has ever seriously studied a poem – known, but set aside in our theoretical doubling – that a word can have a range of signifying functions, literal and figurative, denotative, connotative, systemic, symbolic, and more.

# The Loaded Cat

## Henry Lawson and Jacques Derrida Looking at Animals

*I am I because my little dog knows me.*
 – Gertrude Stein (1936)

*Thus we constantly find ourselves moving in a circle. And this is an indication that we are moving within the realm of philosophy.*
 – Martin Heidegger (1983)

HENRY LAWSON (1867–1922) wrote well over two hundred short stories, some of which rank amongst the best in the world of their time. And many of his stories involve non-human animals. Three of the best known of these are 'The Loaded Dog', in which at least two of the central characters are canine, 'The Drover's Wife', in which, although the story in fact presents us with a panoply of non-human animals, a dog and a snake feature particularly, and 'The Bush Undertaker', which features a large goanna (and which, in the opinion of Inga Clendinnen, 'out-Becketts Beckett at a canter'[78]). In these as in his other stories Lawson's employment of animals is more for humour and the concerns of the narrative than for a concern with animals themselves, although of course there are further dimensions. Dogs are usually present in their capacity as better or worse friends to man. The snake in 'The Drover's Wife' and goanna in 'The Bush Undertaker', however, would seem to emblematise a perhaps uniquely Australian threatening and therefore somewhat Gothic otherness. There is about them – the goanna especially – a touch of the uncanny, as if for Lawson an occasional function of the non-human animal is to look at, from outside, and to some extent ironise, human knowing (much as, as we will find, does a cat in one of the late works of Jacques Derrida).

The goanna is from a family of lizards called *monitor* (the goanna in Lawson's story is very likely a *lace* monitor) supposedly from the capacity of some of them to stand on their hind legs and 'monitor' their surroundings (*monitor*, from the Latin *monere*, 'to warn'). In 'The Bush Undertaker' an old bushman, on an expedition to investigate and plunder – *desecrate* – a native grave he has found, comes across the desiccated body of a friend named Brummy, whom he assumes to have died on his way to visit him. He binds the body between sheets of bark (turns it, as it were, into a sort of *tree*) and drags it back to his hut for 'proper' burial. On the way he notices a large black goanna, and then several more as his journey progresses, and wonders whether there is a sudden infestation of large black goannas in the bush about him, since he's never seen so many before. As I can testify, having more than once as a child encountered what must have been somewhat smaller ones stretched out on our front fly-screen door, the sight of a goanna, one of the largest of lizards, a carnivore, with powerful jaw and claws, can be unsettling – the lace monitor averages one-and-a-half metres in length – and the old bushman is increasingly on edge. At home he props his desiccated friend in a corner and proceeds to comfort himself with drink (a comfort to which we are given to understand he resorts quite frequently; readers might suspect a touch of *delerium tremens* in his vision of multiple goannas), but is drawn out into the night by sounds on his roof. It is the goanna – there has only ever been the one – that has followed him home because he or she has been feeding on Brummy's corpse, and the old man has unknowingly taken off with his/her meal. The goanna throughout the story is *watcher*, and in a sense the old man feels naked before it, perhaps because he is not unaware that there has been something sacrilegious, a violation, about his removal of the Aboriginal bones. Although to some extent Lawson grimly defuses the situation, there's a sense in which (*monere*) the old man is being warned.

We could leave the story there but there is also the matter of Brummy's name. To be *brum* is to come from Birmingham, a town once so infamous for the *groats* counterfeited there that *brum* came also to mean 'false', 'counterfeit' or 'sham'. If we think that the goanna is thus feeding on the sham or false or counterfeit, that the man thus consumed had been/is now dressed as a *tree*, etc., the plot thickens. Did Lawson *intend* us to read the story in this way – the lizard as monitor/devourer of false knowledge, false philosophy (as – as we shall see – Derrida very clearly feels his cat's eyes do)?

Given some metaphoric latitude, there is something of this *watching* function in 'The Loaded Dog' also. The essence of the story is simple. Having at first dragged its fuse through the fire, and thereby igniting it, a young and playful dog runs loose with a sausage-shaped – hence its attraction – explosive device in his jaws, mistaking the way his human friends run away from him for a game and an invitation to chase them. One of those friends runs to the local shanty (here a pub) and the dog follows him, sending the drinkers scurrying. Two of them flee to a second building, the shanty's kitchen. Trying to get to them, the dog runs underneath it only to be chased out again and have his explosive sausage taken from him by 'a vicious yellow mongrel cattle-dog' who has been 'sulking and nursing his nastiness' under the floor. The mean dog settles down with his prize, observed by a dozen others who have suddenly materialised ('spidery, thievish, cold-blooded, kangaroo-dogs, mongrel sheep- and cattle-dogs, vicious black and yellow dogs – that slip after you in the dark, nip your heels, and vanish without explaining – and yapping, yelping small fry') and of course the device at last explodes, and the 'vicious yellow mongrel' is no more.

The men the dog runs amongst are gold prospectors, and while it may be a little fanciful (but texts *do* run away from us, especially when, as in stories or poems, their metaphoric dimensions are amplified) to invoke David Hume's discussion of a *golden mountain* in his *Enquiry Concerning Human Understanding* (1748), or to ask the reader to envision these prospectors as philosophers (or philosoph*ies*), the dog does have, *in his mouth*, the capacity to *explode* his human companions, and the configuration – the underlying ecology – is (implicitly) there.

Although it may interest the reader who does not already know the story to hear that the aforementioned explosive device was originally fashioned in order to harvest fish from a water-hole, and that when it eventually explodes, killing the 'yellow mongrel' (can we divorce this 'yellow' from the racist xenophobia so endemic at the time, especially on the goldfields?), a number of the other dogs are also maimed and traumatised (a small essay could be written concerning the ramifications here), and although one cannot help but reflect that if the dogs here so unflatteringly listed *are* unattractive and ill-tempered it is very likely, since so many of them are 'working' dogs, that it is humans who have made them so, there is perhaps no need to discuss this story further (at least, not just yet). *This* essay, after all, is called 'The Loaded Cat', not 'The Loaded Dog', and I have introduced Lawson's story to some extent as a means of *loading* a discussion of Derrida.

But first there is 'The Drover's Wife', a story about a woman, a slab-and-bark hut – or rather the large, earthen-floored kitchen at the end of it – and a snake. The short third paragraph of this story sets up the situation with admirable concision: 'The drover, an ex-squatter, is away with sheep. His wife and children are left here alone.' Just so. And the story proper – its central action – begins with the sighting of a snake, who disappears first into the woodpile, and then under the house. It is almost evening. The drover's wife – she is not given a name – ushers her children into the kitchen, feeds and puts them to bed, and, with the dog (named 'Alligator'), stays up all night watching for the snake. As she sits awake – the other central action – she goes over in her mind some of the previous trials she has experienced in the Australian bush alone: a bushfire, a flood, a mad bullock, the illness and death of children, threatening and insistent swagmen nosing at her door. 'An evil pair of small, bright bead-like eyes' (*monere*) eventually appears at a crack in the wall between the house and kitchen, and, although sensing danger and attempting immediately to retreat through a neighbouring crack, the snake at last enters, to be caught and killed by the woman and her dog, working together.

In some ways the story follows the same formula as 'The Loaded Dog' and 'The Bush Undertaker'. A house (or kitchen, or shanty) and its human occupants are threatened (invaded, penetrated) by a non-human (dog, goanna, snake). In 'The Drover's Wife' and 'The Loaded Dog' there is, moreover, a further and very significant factor, namely that each story could be said to be *written within a margin of abuse.* In 'The Loaded Dog', the story begins with and is brought about by the intention of creating an explosion in a river so as to kill a large number of fish, and concludes with the death and/or traumatising, similarly on the story's margins, of a number of dogs, and in 'The Drover's Wife' the story begins with and is brought about by the husband-drover's being 'away with sheep', which we might take to mean either his taking them, in search of fodder, from one remote pasturage to another in order to exploit them for their wool, or – more likely at this time – his taking them a long distance or distances overland for slaughter. Can we separate centre from margins? Yes, of course, but the margins will nevertheless intrude, and these stories are about intrusion. Should 'The Bush Undertaker' seem for the moment to stand aside from this further element (the margin of abuse), it is just that: only for the moment. The abuse here – the stealing of the bones – may be of the Indigenous human occupiers of the area, rather than of indigenous non-humans, but the pattern/situation *it* sets up in the story – the sacrilegious stealing of the Indigenous remains *vs* the attempt to *co*nsecrate the remains of Brummy (to give him a 'proper' burial) – merely serves, at a slight intertextual distance, to inflect the structures of the other stories with the same underlay: sacration/desecration.

These stories are also, of course, about *houses*. *Leaky* houses, yes – houses with *cracks* in them, *penetrable* houses – but that is perhaps just their antipodean inflection. We do not need Freud (or Bachelard) to remind us of the house as a metaphor for the mind. We live, as it were, in the *houses of thought* that the mind has made for us. Our philosophies are houses. Whatever else they are about, these stories are also about the (troubled) interrelation of the animal and the human mind – the border-territory, we might say, between the non-human animal and the mind of the human animal. That this may be far from what Henry Lawson might have seen as his intention is hardly the point, as almost a century of literary theory has been trying to tell us.

'The Drover's Wife' ends, melodramatically, with the elder son, Tommy, declaring to his mother, as they watch the 'mangled reptile' burn in the fireplace: 'Mother, I won't never go drovin'; blarst me if I do!' From a sheep's perspective, it would be nice to think this registered an animal turn, but of course it's nothing of the kind. To suggest the story is entirely unambivalent in this regard, however, would be wrong.

Earlier in the story, as the children settle down to sleep on the kitchen table, Tommy is given to say 'Mother! Do you think they'll ever extricate the (adjective) kangaroo?' Idle chatter, it may be, but, from the author of 'The Bush Undertaker', we might be inclined to ask why the eradication of the kangaroo should be mentioned at all, if we are not being asked to reflect upon it, look at the malapropism more closely. 'Extricate'? 'To free someone or some thing from a constraint or difficulty'? Within the story, that is to say, there could be seen to be something like a hole, a trap-door, into some other space entirely. However else we might see it, the kangaroo is certainly enmeshed, entangled, disastrously in the Western mind. But that, perhaps, is another story.

It is time we turned to the cat.

SHE COMES TO US – WE ARE TOLD emphatically this cat is a 'she' – as a character in a late book of Derrida's. The book is entitled *The Animal That Therefore I Am* (*L'animal que donc je suis*), a title so intricate that simply by exploring its nuances one could offer a precis of human/animal relations in philosophy since Descartes, upon whose famous *Cogito ergo sum* of course it leans. The book is built up from Derrida's notes for, and transcriptions of, a set of long lectures he gave to a conference on his work (*L'animal autobiographique*) at Cérisy la Salle in 1997. Although the book as a whole was published posthumously (in 2008: Derrida died in 2004), its first section, itself entitled 'The Animal That Therefore I Am', was published in French in 1999, and in English translation in 2002, in a form edited and authorised by the philosopher himself.

The cat appears almost immediately, as if Derrida has been keen to tell us about her. When we 'meet' her, she is sitting in his bathroom – locus of abjection, site of preparing-oneself-for-the-world – observing the philosopher naked. Which makes the philosopher, *as* philosopher, *feel* naked, an experience, perhaps, not unlike that of the Emperor in Hans Christian Andersen's 'The Emperor's New Clothes', when the young boy in the crowd, uninhibited by the fear and self-doubt that silences the rest of the crowd, declares that he, the emperor, is parading before them clothed in nothing at all. Before the gaze of the non-human animal – such is Derrida's assertion – the philosopher is embarrassed, his philosophising no longer able to clothe him. Embarrassed, and also in some measure apprehensive, since we are given quite clearly to understand that the focus of the cat's attention is Derrida's penis. A *castration anxiety* is therefore involved. Were it anyone else but Derrida we might think that that was all it was and, perhaps a little abashed that this information has been shared with us, move on. But of course this goes to the core of the matter. We are talking – or perhaps it would be better to say we cannot be sure that we are *not* talking – obliquely, of phallogocentricity, and the challenge, the threat of castration, that 'the Animal'[1] presents to it.

But how appropriate is his choice of the cat as the vehicle of this perception? A creature – a species – that has not only accompanied the human, has been *domesticated,* for thousands of years, but that has such a presence in folklore, mythology, literature, even philosophy, that it is hard to see how it can be in any way an innocent player. The cat, we could say at the outset, is *loaded,* so heavily in this way that one wonders whether she can, in truth, be seen as anything other than *mirror.* She may, that is, sit Sphinx-like observing him, her stance as it were unclothing his philosophy, but (and I must add that Derrida nowhere mentions the Sphinx) she *is* Sphinx-like – few readers will not find such a thought passing through their minds (i.e. *themselves* loading the cat) – and the Sphinx, we might reflect, was not only half-lion/half-human (already non-human animal and not), but presented Oedipus most famously with a riddle (What goes on four legs, then on two legs, then on three?) the answer to which was Man.

I would wonder, too – Derrida is Francophone after all – if there wasn't, in any writing he might do about a cat, more than the ghost of a memory of Baudelaire's famous cat poems, 'Le Chat' and 'Les Chats', the latter of which was the subject of one of the key essays (by Roman Jakobson and Claude Lévi-Strauss, 1962) in structuralist analysis.[2]

---

1   A term I will use in Derrida's despite – *pace*, that is, his extensive argument, in the book concerned, regarding the term 'animal' as an intellectual violence. See footnote 1 in the chapter 'Animal Dreams'.

2   Available variously, for example, in Roman Jakobson, *Language in Literature,* ed. Krystyna Pomorska and Stephen Rudy (Cambridge, MA: Harvard University Press, 1987), 180–97.

If it is hard to see how Derrida could write about a cat without the Sphinx, or Baudelaire, or Jakobson and Lévi-Strauss in mind, it is almost impossible to imagine that he could do so without some thought to Maurice Blanchot. Blanchot's early uses of the cat – or, rather, the texts in which those uses occur, but one must emphasise that these uses are central to these texts, as well as to the definition of Blanchot's own relation to 'the Animal' – are of considerable significance to the subsequent development of literary theory and its intensification of the interrelation of literature and philosophy. Blanchot himself was an important influence in the development of Derrida's thought, became eventually a friend, and was the subject, indeed nominal co-author, of one of Derrida's last books.[3]

'Everyday language', writes Blanchot in 'Literature and the Right to Death' (1947):

> calls a cat a cat, as if the living cat and its name were identical, as if it were not true that when we name the cat, we retain nothing of it but its absence, what it is not. Yet for a moment everyday language is right, in that even if the word excludes the existence of what it designates, it still refers to it through the thing's nonexistence, which has become its essence. To name the cat is, if you like, to make it into a non-cat, a cat that has ceased to exist, has ceased to be a living cat, but this does not mean that one is making it into a dog, or even a non-dog.

– lines which could be seen as themselves an explanation of a pivotal passage in the revised version of his first novel, *Thomas the Obscure*, in which Thomas encounters, late at night, a nearly-blind cat who, following him through the house and who, 'slipping into a tunnel where he did not recognise a single smell ... filled his lungs and howled',

---

3    Maurice Blanchot/Jacques Derrida, *The Moment of My Death/Demeure*, trans. Elizabeth Rottenberg (Stanford: Stanford University Press, 2000). Derrida's text was originally published as *Demeure: Maurice Blanchot* (Paris: Galilée, 1998). Kevin Hart, in personal correspondence, cautions that to present the two works together was a publisher's decision and should not be taken to evidence any particular personal relation between the authors.

bemoaning that not only have his usual cat spirits ('the spirit that tugs at my tail when the bowl is full, the spirit that gets me up in the morning and puts me to bed in a soft comforter') have abandoned him, but that, 'surrounded by a special void which repels me and which I wouldn't know how to cross over', he has been horribly transformed:

> My body, which is just like the body of a man, the body of the blessed, had kept its dimensions, but my head is enormous ... and rather than a head seems nothing but a glance, just what is it? ... I am dead, dead. This head, my head, no longer even sees me, because I am annihilated. For it is I looking at myself and not perceiving myself.[4]

Quite simply, he has strayed into a novel; strayed, or been drawn, into *language*, into *text*. Deprived thus of his *Dasein*, his 'actual' being, he finds himself, *in the realm of the human*, at once in a particular/peculiar *emptiness*, and taking on (Sphinx-like) *human* form, since the form of cat per se is impossible, is annihilated in that place.

One wonders whether, given a chance to respond, *allowed to get a word in edgewise*, in that *impossible* place of response, Derrida's cat might say something similar, that is, that in Derrida's text she is not *seen*, has taken on *human* form, has become, as it were, a mirror. We should perhaps try to leave aside for the moment the intriguing question of what it might mean for Derrida, master of the trace, of *différence*, he who says that there is *nothing that is not text*, to try to remove this cat from such a situation by proclaiming her *reality* – although we should at the same time recognise that it is a paramount question. Indeed, Derrida himself seems somewhat shocked, as if realising that he has just attempted to strip the cat naked of all

---

4    It is generally held that Blanchot wrote *Thomas l'Obscur*, originally a book of over two hundred pages, between 1932 and 1940. First published in 1941, it was in the late 1940s (finished 1948) extensively revised and shortened to approximately one-third of its original length. Published in 1950, this revised version was translated by Robert Lamberton and published in 1973. My quotation here is from *The Station Hill Blanchot Reader*, ed. George Quasha (Barrytown, NY: Station Hill Press, 1999), in which Lamberton's translation is reproduced on pages 51–128.

association, *to draw her out of the cloud of language,* against so much that he has argued before.

Derrida himself, to acknowledge this at the outset, seems conscious of this seeming paradox, and of the loading which attends it, and tries, in the very act of introducing his cat, to fend some of it off. 'I must make it clear from the start,' he says,

> the cat I am talking about is a *real* cat, truly, believe me, *a little cat.* It isn't the *figure* of a cat. It doesn't silently enter the room as an allegory for all the cats on the earth, the felines that traverse myths and religions, literature and fables. (374/6)

Indeed he seems so nervous on this point (or is he merely tongue-in-cheek?) that he repeats it in the next paragraph. 'An animal looks at me. What should I think of this sentence? The cat that looks at me naked and that is *truly a little cat, this* cat which I am talking about, which is also a female, isn't Montaigne's cat':

> Nor does the cat that looks at me naked, she and no other, the one I am talking about here, belong, although we are getting warmer, to Baudelaire's family of cats, or Rilke's, or Buber's. Literally speaking at least, these poets' and philosophers' cats don't speak. 'My' pussycat (but a pussycat never belongs) is not even the one who speaks in *Alice in Wonderland.* Of course, if you insist at all costs on suspecting me of perversity – always a possibility – you are free to understand or receive the emphasis I just made regarding 'really a little cat' as a quote from chapter 11 of *Through the Looking Glass.* Entitled 'Waking,' this penultimate chapter consists of a single sentence: 'and it really was a kitten, after all' ...
>
> Although time prevents it, I would of course have liked to inscribe my whole talk within a reading of Lewis Carroll. In fact you can't be certain that I am not doing that, for better or for worse, silently, unconsciously, or without your knowing. (376/7)

Ah, so there we have it, not only an admission that the cats of Montaigne, Baudelaire, Rilke, Buber, and doubtless many others

(Blanchot) *have* been in his mind, but that one cat, a kitten in fact, from *Through the Looking Glass*, has been especially so.

What to make of this? An admission that the mirror *is* a problem? That, ideally, one should be able to pass *through* that mirror? An admission, since this *is* a reference to Alice's *waking*, that the texted/textual world *is* like a dream, from which there might, just, somehow, *be* a waking into a world of 'reality' (Rilke's, Heidegger's 'Open', let's say)? Perhaps, though one must also remember that Alice woke in her own armchair (holding the black kitten), on the original side of the mirror, with an implication that she had never passed through at all.

The trembling that we see here, the almost-admission of a need or desire to grasp a *real*, a mode of Being unshackled by a consciousness of its relentless textuality, characterises the whole of *The Animal That Therefore I Am*. Derrida asserts that the cat of which he speaks is herself alone, that she is a '*real*' cat, that she is not 'the figure of a cat', and then admits that even the statement that she is just a kitten is a quotation, as if the whole has been a game, a gambit, to demonstrate the point that he has been labouring for years, that there is nothing that is not – that there is no way of escaping – text. But the mere fact that he would exercise this gambit, play this game at all, seems to betoken a discomfort, a heightened consciousness – before the stare of this cat – of a constraint, an inadequacy, an impairment, in this very position. When later in the essay he speaks directly about the horror of factory farming and our industrialisation of non-human animals, a process he readily compares to the *Shoah* –

No one can deny seriously any more, or for very long, that men do all they can in order to dissimulate this cruelty or to hide it from themselves; in order to organize on a global scale the forgetting or misunderstanding of this violence that some would compare to the worst cases of genocide (there are also animal genocides: the number of species endangered because of man takes one's breath away).[5]

---

5     *The Animal That Therefore I Am*, 25/26. 'One should neither abuse the figure of genocide', the passage continues, 'too quickly consider it explained away … [T]he annihilation of certain species is indeed in progress, but it is occurring

– it is all too briefly, and swiftly followed – as if he has shocked even himself, is speaking in a mode he simply cannot sustain – by a return to more guarded, 'philosophical' discussion.[6] There is deferment. We are promised an eventual and extensive examination of philosophy's relation to 'the Animal'. And when this at last comes, in the form of *The Beast and the Sovereign*, it is itself a kind of deferment.

What has happened? Has Derrida reached, embarrassed, the limits of (his) philosophy? Has he offered himself up as an example of its failure? Has he *stumbled*, in the face of non-human animals? Has he opened this door at last, this great door, just a crack, and looked through, to find the territory so vast, so strange, that he feels naked before it, and has to close it, and back away? I suspect that it is all of these things, if these things are separate at all. Or rather, to put this in another way, that Derrida has realised that, behind the mirror-cat, which he, and a long tradition behind him, has so loaded that he cannot see past it, there is another cat, another *being*, loaded with itself, its suffering, the weight and intensity of its own existence. An *abyss*, in short, which threatens to devour him.

---

through the organisation and exploitation of an artificial, infernal, virtually interminable survival, in conditions that previous generations would have judged monstrous, outside of every presumed norm of a life proper to animals that are thus exterminated by means of their continued existence or even their overpopulation. As if, for example, instead of throwing a people into ovens and gas chambers (let's say Nazi) doctors and geneticists had decided to organize the overproduction and overgeneration of Jews, gypsies and homosexuals by means of artificial insemination, so that, being continually more numerous and better fed, they could be destined in always increasing numbers for the same hell, that of the imposition of genetic experimentation, or extermination by gas or by fire.'

6   I am not the first to articulate disappointment at the *volte-face* that circumscribes so quickly Derrida's much-vaunted 'animal turn'. Haraway (2008), for example, writing of the philosopher's cat scene, speaks of his failing 'a simple obligation of companion species': 'he did not become curious about what the cat might actually be doing, feeling, thinking, or perhaps making available to him in looking back at him ... Incurious, he missed a possible invitation, a possible introduction to other-worlding. Or, if he was curious when he first really noticed his cat looking at him ... he arrested that lure ... with the sort of critical gesture that he would never have allowed to stop him in his canonical philosophical reading.' (20)

# Field's Kangaroo

*What manner of animal is this? ... This dog-faced bear who has swallowed a python?*
    – John A. Scott (2014)

SPARE A THOUGHT for Barron Field's 'The Kangaroo', a most intriguing poem. Not, perhaps, the first written on Australian soil, but one of two poems in what by all accounts was the first book of poetry published in the colony of New South Wales, a slim volume called *First Fruits of Australian Poetry*, printed by the government printer, George Howe, in 1819, and subsequently republished by his son Robert (George having since died of the dropsy) in an expanded edition in 1823.

Although it has occasionally found defenders, 'The Kangaroo' has been much maligned,[1] largely as a poor attempt to adapt English Romanticism to an Australian subject, and the desire to display it in anthologies of Australian poetry has come and gone. Mitchell and Kramer, for example, in their *Oxford Anthology of Australian Literature*

---

1   'Poor are the first fruits of a Barron Field,' wrote Edward Smith Hall, another colonial bard, quite nastily, in 1828, 'To human industry is wont to yield; – / Poor are the first fruits of thy sterile brain, / Conceiv'd by folly, and brought forth with pain!', and it has gone on: 'poetically slight' (Michael Ackland); 'comic' (Paul Kane).

(1985), chose to pass over it, as did Gray and Lehmann in *Australian Poetry Since 1788* (2011). Although the poem does contain some awkward lines, its principal crime – and embarrassment – seems to have been to rhyme 'Australia' with 'failure', though on the one hand it's hard to see as a crime and embarrassment a rhyme so often and so ruefully quoted, as if it contained a deep, sad truth, and on the other hand the understanding that the poem is actually *saying* this – that is, substantiating a connection between Australia and failure – is misleading. Indeed, this pre-emptive and adventitious misreading – the poem, generally speaking, has been subjected to *raiding* rather than reading – usurps the poem at the outset and denies us a curious poetic jewel: a rough diamond, admittedly, but diamond nevertheless. Field is in fact saying something quite different, and in its intellectual engagement with its subject the poem represents a rather extraordinary achievement. In its struggle with the *concept* of the kangaroo, we get an index of the whole initial and much-protracted intellectual encounter with Australia, in short the eighteenth- and nineteenth-century Western mind's encounter with its other. More than enough, surely, to earn the poem a permanent place in the rather bleak, troubled, and betimes quite Gothic dawn of Australian writing.

But where to start? Perhaps with the poem's own dawn: its epigraph, *mixtumque genus prolesque biformis*, from book six of the *Aeneid*. While this line might have been chosen for other reasons – it's in book six, for example, that Aeneas descends to the Underworld (Australia = bottom of the world = Underworld) – Aeneas is never a casual reference in English poetry. Aeneas settled in Italy after the fall of his native Troy. His son would become one of the founders of Rome, and, according to the ninth-century *Historia Brittonum*, his grandson, Brutus, became the founder/discoverer of Britain (its name derived from his own), at that time a wild and desolate place on the westernmost edge of the world, inhabited by giants. The antipodean ring here (Abel Tasman's surmise, when he saw cuts in the trees that he took to be footholds, that the people of Van Diemen's Land must be giants) is only confirmed when we consider that the *Aeneid*'s author is Virgil, chosen by Dante to be his guide through Hell and Purgatory (after all, Virgil had been there before), and so to explain, as the pair emerge from one into the other, having just walked along the spine of

Lucifer, that they are now in the Antipodes. If the world is a giant man, Virgil might have been saying, we are now where the feet are.

Giant feet. *Macropoda*. Field's poem is off and bounding before its own first line has started. As to the actual text of the epigraph, that translates as 'the two-formed offspring of a blended birth', and refers (ostensibly) to the Minotaur, the hybrid progeny – head of a bull and body of a man – of Pasiphae, wife of King Minos, and the white bull of Poseidon. This, too, is a rich summoning on Field's part. One could mention a few stories here – how Minos had asked Poseidon for help in defeating his brothers for control of Crete; how Poseidon had sent him a mighty bull on the understanding it would eventually be sacrificed in his honour; how Minos, cheating, had kept that bull and sacrificed another; how, to punish Minos, Aphrodite had made Pasiphae fall in love with the bull; or how Pasiphae had the royal craftsman, Daedalus, make a wooden cow into which she could climb, to experience the white bull's potency. Daedalus, subsequently, built the labyrinth in order to imprison their offspring. But we could go to Virgil for this – it's all there, in book six of the *Aeneid*. Field's poem, as we shall see, will present its own menagerie.

'Kangaroo, Kangaroo!', it begins, 'Thou Spirit of Australia', and we could pause right there, to point out, say, that in the late 1970s Ken Warby built, in a Sydney backyard, the wooden speedboat with which he broke the world water speed record and which he called the *Spirit of Australia*; that Australia's national airline, Qantas, dubbed 'the flying kangaroo' for the motif carried on the tail of its planes, has also styled itself 'The Spirit of Australia'; or, more pertinently, that Field himself is thus, at the very outset, presenting the kangaroo as emblematic of its wider place, encouraging the reader to see some of the things he says about the kangaroo as about Australia itself.

A more pertinent point is that already, in his second line, Field is making a bolder and more topical statement than we, two hundred years on, are likely to realise. Matthew Flinders, in a voyage of 1802–3, had become the first to circumnavigate the Great South Land. He chronicled his expedition in *A Voyage to Terra Australis*, a book in which, in a passage discussing the name 'Terra Australis', he makes the following suggestion:

Had I permitted myself any innovation upon the original term, it would have been to convert it into Australia; as being more agreeable to the ear, and as an assimilation to the names of the other great portions of the earth.

Flinders' book was published in 1814, his plan to write up his voyage having been long delayed by his arrest, as a suspected spy, on the French island of Mauritius. It was a further three years – 1817 – before Governor Macquarie received a copy and began to use the term 'Australia' in official correspondence. Field's very use of the term 'Australia', this is to say, is arguably the first in poetry anywhere.

Let's look at the first stanza in its entirety:

> Kangaroo, kangaroo!
> Thou Spirit of Australia,
> That redeems from utter failure,
> From perfect desolation,
> And warrants the creation
> Of this fifth part of the earth,
> Which would seem an after-birth,
> Not conceiv'd in the Beginning
> (For GOD bless'd his work at first,
>     And saw that it was good),
> But emerg'd at the first sinning,
> When the ground was therefore curst; –
>     And hence this barren wood!

Straight away we have that notorious rhyme. And straight away, too, we can see it's not so simple. The lines are trying to weaken, not strengthen, a connection between Australia and failure. The kangaroo *redeems* the land from 'perfect desolation' and in fact *licenses* its creation. But negative suggestion is also at work. In mentioning 'failure' and 'desolation', even if only to deflect them, the poem places them in the reader's mind, makes disconnection and connection simultaneously. Indeed it's stronger than this. The words 'utter' and 'perfect' make the statement comparative. The kangaroo redeems Australia from *utter* failure, from *perfect* desolation, but the failure and the desolation remain. Ambivalence. That Aussie tic. Right at the start. This whole stanza is quite literally *riddled* with it.

The 'That' in the third line, for example. We might almost automatically assume it to refer to the kangaroo, but, grammatically, it might also refer to 'Australia', or perhaps to the broader '*Spirit* of Australia', in which case, as if we had turned an opal slightly, we might have revealed a rather different set of questions, such as *what* or *whom* Australia redeems from utter failure and desolation, and what conception of 'Australia' we might now be looking at. For here there would seem to be, hovering in the background, that sense of Australia as a possible new Eden that has so long been an aspect of Australian literature, a chance to get right, in this new world (a Dantescan purgatory?), the things that went so wrong in the old. The word 'redeems', that is to say, might have a deeper resonance. But to fathom it we must read on.

Even in the twenty-first century there is no universal agreement upon the number of continents. Australia is the sixth or seventh, depending upon how one counts the rest. In 1819 it was probably regarded as the fifth, since the first confirmed sighting of Antarctica (by the expedition of von Bellingshausen and Lazarev) did not occur until the year after Field's poem was published. In referring to Australia as 'this fifth part of the earth', then, line six might simply be citing what was then understood to be geographical fact. But it might also allude to ancient and enduring senses of the world as made up of four cardinal points – four corners – or four elements (earth, fire, wind, water), in which case a 'fifth part' would be quite literally *out of thought*, not part of the earth's original conception. And indeed this appears the sense of the stanza. The continent thus appears 'an after-birth', and here again there is a fork in meaning: an after-birth as in something born afterwards ('Not *conceiv'd* in the Beginning'), but also the placenta, which, in the process of birth, is delivered after the child – a reference to something which, while not part of the earth as originally conceived, nevertheless has sustained or nourished it. A mixture, here – perhaps a confusion – of mortal and immortal processes. The earth both born *and* created. Clearly we are not out of the woods yet (in fact we're just approaching them!).

'GOD', that is, *conceiv'd* the world in its original, four-cornered form, 'And saw that it was good'. He did not think about a fifth part until Adam and Eve had committed their original sin, in eating of the Tree of Knowledge. This sin – embarking upon a process of knowing – occasions their exile. And exile in turn necessitates not only a place to be exiled *to*, but that that place be *un*-Eden-like, 'curst' (a *penal colony?*). We are a long way before Derrida's articulation of the *supplement*, but there is surely something of it here, in the representation of the fifth part, in the relation of the after-birth to the born, and in the epistemological cast of diction ('conceiv'd' as a mental as well as a physical act) and image (the nature of the first sin). Australia, and the kangaroo as its avatar, are presented as *outside received thought*, something happening to the *mind*.

'And hence' – another intriguing twist – 'this barren wood!', a phrase which most immediately signifies the land of desolation encountered in the opening lines of the poem but which surely should give even a casual reader some pause and amusement. This poem is written by a 'Barron Field' ('Barron' was his mother's maiden name), and here we are not only given a 'barren wood', but an exclamation mark for good measure. Is this a signature? Has the poet thrown himself into the equation? Why? Could it be he who has somehow – also – been redeemed?

Perhaps 'wood' gives us a clue. It's a key word, after all, in some of the most famous lines in Western literature, the opening of Dante's *Commedia* (here in Dorothy Sayers' translation), that great poem of exile:

> Midway this way of life we're bound upon,
> I woke to find myself in a dark wood,
> Where the right road was wholly lost and gone,
>
> Ay me! how hard to speak of it – that rude
> And rough and stubborn forest!

– and I am not sure that the original of these lines isn't in Field's mind here. In 1817, when he arrived in New South Wales, Field (1786–1846) was truly in the middle of his life, although there's no way he could have known it. And there's much in his other writings to suggest that he felt, at least for a time, in exile. Am I pressing too hard for an allusion here? Perhaps, but, first, the motif of exile will return, and second, such subtle allusion (by the knight's move, I call it: two steps forward, one to the side) would not be unique in this poem. Who would have thought, for example, that the opening lines of the third stanza –

> She had made the squirrel fragile;
> She had made the bounding hart;
> But a third so strong and agile
> Was beyond ev'n Nature's art;
> So she join'd the former two
> In thee, Kangaroo!

– would in fact be, as John Byrnes points out, a very clever recasting of Dryden's epigram on Milton:

> Three poets,[2] in three distant ages born,
> Greece, Italy, and England did adorn,
> The first in loftiness of thought surpassed,
> The next in majesty, in both the last:
> The force of nature could no further go;
> To make a third she joined the former two.

But that, indicative as it may be of how Field's mind works, is to jump ahead.

At the heart of that first stanza, in the middle of its middle line, is the word 'seem'. The whole stanza turns upon it. The desolation, the failure, the barrenness, are – or may be – apparent only. This doesn't mean the ambivalence is resolved, but it is reframed, shifted. The second stanza takes up immediately this matter of appearances: '*Tho*' at *first sight*', it tells us, we might say that in the kangaroo's nature 'there may / Contradiction be involv'd', this is only an illusion and 'like discord well resolv'd / It is quickly harmoniz'd'. The *deep mental action* of this

---

2   Homer, Virgil (again) and Milton.

poem – as the first stanza foreshadowed, with the need to supplement the four-square earth with a fifth part, the need for God, who had 'bless'd His work at first', to make a correction to it – is about *coming to be*, a process, at once of imagination and of realisation, neatly caught in the next lines:[3]

> Sphynx or mermaid realiz'd,
> Or centaur unfabulous,
> Would scarce be more prodigious,
> Or labyrinthine Minotaur,
> With which great Theseus did war,
> Or Pegasus poetical,
> Or hippogriff – chimeras all!

'At first sight' this may seem little more than a list of mythic creatures, but an argument runs through them: were the Sphynx or the mermaid to be brought into the realm of reality, or a centaur to prove not just a creature of fable, they – like the minotaur in its labyrinth, the winged horse that has become a figure of poetry itself, or its cousin-in-imagination the hippogriff, hybrid of horse and the already-hybrid griffon (body of a lion, head and wings of an eagle) – would not be more remarkable than the kangaroo, which *does* exist. Becoming, as I say. Emerging. Into a reality hitherto only imagined.

At the end of this rather Borgesean list of imaginary beings, moreover, Field adds 'chimeras all!', at once a dismissal of the list and an addition to it, as if in afterthought, or there were something that, in this transition from imagination to reality, Field wants to make sure we carry across. A chimera is a wild fantasy, an impossible illusion, but also, of course, herself a mythic creature, another hybrid (head and body of a lion, tail that becomes a snake, head and neck of a goat arising from the middle of her back), slain, eventually, by Bellerophon, who was able to achieve this because he was riding Pegasus, and thus able to attack her from the air, out of reach of her heads and fiery breath

---

3    I am observing Field's addition, to his second edition of *First Fruits*, of two lines about the minotaur, absent from most anthologised versions of the poem.

(poetry assisting in the slaying of illusion?). She was also – a point which Field does not use, though it seems to beg to be here[4] – a child of Echidna, 'mother of monsters', herself half-woman, half-snake.

Even the mermaid has her by-story. Field's poem might not talk about the *Australian* echidna (except very obliquely), but it ends with mention of her biological cousin the platypus, or, if one chooses to see it that way, with the footnote, after its last line, concerning the platypus ('The *cygnus niger* of Juvenal', that footnote reads, 'is no *rara avis* in Australia; and time has here given ample proof of the *ornythorinchus paradoxus*'). Field, of course, does not *call* the platypus a platypus. He chooses, instead, to employ the term 'duck-mole', in currency at the time, perhaps to emphasise the motif of hybridity. When evidence of the platypus (*ornythorinchus paradoxus*) was first seen in England – a sketch, and a pelt (in a stinking barrel) sent by Governor Hunter in 1798 – it was greeted with considerable suspicion. Exotic animal forgeries had been coming from dubious taxidermists in the Far East for centuries. One favourite, apparently, was the mermaid made of the head of a monkey and the tail of a fish, artfully sewn together. The platypus, it was thought, with the bill of a duck, the tail of a beaver, the feet of an otter, was one such ingenious construction. Field would have known of these suspicions. The colony itself would have been rife with such stories, and very likely he had read George Shaw's *Naturalist's Miscellany* (1799), in which the eminent naturalist, writing of Hunter's specimen, stated that it was 'impossible not to entertain some doubts as to the genuine nature of the animal'.

Four of the imaginary creatures we've just met – the Sphynx, the mermaid, the centaur and the minotaur – are human/non-human animal hybrids, and of these at least two are allied with wisdom: the Sphynx, who asked the riddle at the gate of Thebes, and the centaur, who through the figure of Chiron is associated with teaching. But Chiron was an exception. The centaurs are just as well known for lustfulness. Given to intoxication and violence, they symbolise our animal nature – or, rather, the animal nature from which we are supposed to have lifted ourselves. It is interesting that Theseus figures

---

4    A point not lost on a later poet, A.D. Hope, whose 'The Drifting Continent' turns upon it.

in their story, as a warrior central to the victory of the centaurs over the Lapiths (a tribe of humans, from Thessaly), since Theseus, too, figures large as the slayer of the minotaur, who could himself be seen as an embodiment of man's divided, civilised/animal nature – the labyrinth, from this vantage, symbolising our need to repress or imprison the animal in ourselves, and civilisation, through the figure of Theseus ('maker of cities'), a victory over our baser nature.

What does this have to do with the poem? We have, first, in the human/animal hybrids, an association of the human with the animal, or evocation of the animal *in* the human, a connection we might at first think a thing to be avoided. And we have, second, an association of the animal/human hybrid with wisdom – something (wisdom) with which we shall find Field somewhat preoccupied. 'But what Nature would compile,' Field now tells us, taking up the earlier lines about discord harmonised in the kangaroo, 'Nature knows to reconcile; / And Wisdom, ever at her side, / Of all her children's justified.' Wisdom on the side of nature? Wisdom outside or beyond the *mythos* of the hybrid and imaginary creatures we've just had gathered for us? Wisdom associated with the Antipodes? And with an animal – 'the Animal'? Another ambivalence? We'll see.

We might say that this poem is strung upon four processes of becoming, one a movement from the Western mind into the antipodean, a second from a divine explanation of creation to an explanation which not only accommodates a strong 'natural' component but sees that component as completing a story, harmonising an earlier discord, a third, which we might describe as a deepening encounter with the Animal, and a fourth which we might see as a movement from imagination into reality, as if imagination had somehow been reality's antechamber – as if, indeed, imagination *brings things about*, can be a harbinger of a new shape of reality. And of course none of these are mutually exclusive. A wider 'reality' is approached as one draws away from the gravitational field of the Western mind. One is led towards this more open knowing by an animal – the kangaroo – who itself stands outside one's original understanding of being. The 'natural' supplements and challenges, or perhaps (for Field is a good Christian) merely clarifies, the original story of creation. This thinking is still four decades before Darwin's great work appears, but from our own vantage point we can see him fairly clearly on the horizon. I don't think we should be too surprised if Field's thinking seems groping and unclear – just about everyone's was, on these issues – but maybe it's a little clearer than it at first appears.

'She had made the squirrel fragile', the third stanza begins:

> She had made the bounding hart;
> But a third so strong and agile
> Was beyond ev'n Nature's art;
> So she join'd the former two
>    In thee, Kangaroo!
> To describe thee, it is hard:
> Converse of the camélopard,
> Which beginneth camel-wise,
> But endeth of the panther size,
> Thy fore half, it would appear,
> Had belong'd to some 'small deer,'
> Such as liveth in a tree;
> By thy hinder, thou should'st be
> A large animal of chace,
> Bounding o'er the forest's space; –
> Join'd by some divine mistake,
> None but Nature's hand can make –
> Nature, in her wisdom's play,
> On Creation's holiday.

It pays to register that the *deer*, introduced by the 'bounding *hart*' of the second line, plays through the stanza that follows. We see it in 'small deer', we see it in the '*hind*er' parts of the 'large animal of chace' (surely a *hind*). And as for 'To describe thee, it is hard': can we dismiss the possibility that it sacrifices a measure of grace in order to catch an echo? The fourth and fifth lines of Dante's *Commedia* – I've already spoken of the first three – read 'Ay me! how hard to speak of it – that rude / And rough and stubborn forest!'

It's a push, I'll readily admit, this Dante business: a long bow. There is, for a start, the question of translation (although how many possibilities are there for *Ahi quanto a dir qual era è cosa dura / esta selva selvaggia*?), let alone the fact that the one most readily available to Field (Henry Francis Cary's, 1805) does *not* use 'hard'. But in support is the curious fact that 'hard', in the Field, rhymes with 'pard' (or 'leopard', depending upon the version consulted) and that, at just this point in the *Commedia*, Dante himself, having tried to escape the dark wood, finds his way blocked by a leopard. Coincidence? Perhaps. But Field, as we've seen already, is a logodaedalist. We might choose to find support in another and more overt allusion, this time to Shakespeare. 'Small deer' may mean just that, a small deer, but it's given us in inverted commas – Field wants us to know that he's quoting – and in fact leads us to a line from *King Lear*, 'Mice and rats and such small deer', a reference to the diet of Tom O'Bedlam. And Tom O'Bedlam, of course, is Edgar, banished, exiled, just as was Dante, and just as it seems Field sees himself to be. Things in exile. Things in hiding. Things not as they seem. A personal tale, perhaps, woven through the poem like a dark thread.

Allusion aside, the stanza is most immediately concerned, as was the previous, with the 'construction' of the kangaroo. Nature, having exhausted herself with the creation of the squirrel and the hart, and unable to come up with a third 'so strong and agile', has created her own hybrid, and combined the former two. 'To describe thee, it is hard;' – to take up at that awkward line – 'Converse of the camélopard'. Another hybrid, a combination of the camel and the leopard. But not a hybrid in the mythic sense of the minotaur, griffon or chimera. This hybridity is *descriptive*. The giraffe (even now it is *Giraffa camelopardalis*) *looks* (although perhaps more so to someone who hasn't actually *seen* a camel) like a camel with a leopard's spots. The possibility notwithstanding that the kangaroo is 'converse' of the camelopard in that it begins with a small head but has large hindquarters, it is also *converse* in that other sense (we should have learned by now that, with Field, not all words are as they seem), of familiar or cohabitant with: like the camelopard, the kangaroo is a creature that *does* exist, that is hybrid of other creatures that *do* exist.

Should we doubt that the use of 'converse' may be this less frequent one, we have, as if in a kind of linguistic confirmation, the strange construction upon which this stanza finishes:

> Join'd by some divine mistake,
> None but Nature's hand can make –
> Nature, in her wisdom's play,
> On Creation's holiday.

These lines make little sense, even seem contradictory. How – unless, of course, the 'divine' and 'Nature' are one – can a 'divine mistake' be made by 'None but Nature's hand'? And haven't we earlier been told that 'what Nature would compile, / Nature knows to reconcile'? If, however, we read 'Join'd' in a similarly archaic sense, as *enjoined*, the gist changes: 'prompted' or 'urged' by 'some divine mistake', only Nature is able to create – a reading quite consonant with the two lines that follow. We have been thrown out of Eden, and, in this fallen world, the world of the *mistake*, it is Nature's task to continue the processes of Creation.

And so the poem progresses towards its end, an accommodation that sees the kangaroo, strange as it might at first have seemed, and Australia with it, as congruent with, rather than antipathetic to, existing creation. 'For howso'er anomalous', the fourth stanza begins:

> Thou yet art not incongruous,
> Repugnant or preposterous.
> Better-proportion'd animal,
> More graceful or ethereal,
> Was never follow'd by the hound,
> With fifty steps to thy one bound.
> Thou can'st not be amended: no;
> Be as thou art; thou best art so.

– the image fulfilling the earlier promise of 'a third so strong and agile', in giving us an animal at once more graceful and more powerful than the deer, and (a suggestion that the kangaroo will elude our sense, our understanding still?) more suited to *out-run the hounds* of reason, understanding or taxonomy that would pursue it.

'When sooty swans are once more rare' reads the first line of the final stanza:

And duck-moles the Museum's care,
Be still the glory of this land,
Happiest Work of finest Hand!

We know, now, that Work and that Hand to be Nature's. But the territory of the Supplement (as Derrida tells us in '... That Dangerous Supplement ...') is a tricky one. Nature might provide a supplement to a Creation – or a Creator – caught out by the unexpected sin of those to whom he had given dominion, but it is hard to imagine on the one hand that that sin *was* unexpected, or on the other that the Supplement was not already part of Creation. The ambivalence we encountered at the beginning of the poem seems to have continued. Just as the previous stanza, in attributing to the kangaroo a kind of finished perfection, also intimated its out-running the hounds that would pursue it, so Field himself now offers us, through a footnote, a supplement of his own, at once completing his poem and showing that it is *in*complete. 'The *cygnus niger* of Juvenal' – to quote it again – 'is no *rara avis* in Australia; and time has given ample proof of the *ornithorhynchus paradoxus*'.

Juvenal, in his sixth satire, (in)famous for its misogyny, had written that the truly beautiful and virtuous wife was as rare a bird (*rara avis*) – he meant impossible – as the black swan: *rara avis in terris nigroque simillima cycno* ('a rare bird in the earth and most similar to a black swan'). And yet, of course, the black swan – and by implication the beautiful and virtuous – is an actuality, and no rarity, in Australia. So too, by 1819, when Field writes his poem, time has well demonstrated the authenticity of the platypus (*ornithorhynchus paradoxus*: genus 'bird-snout'; species 'paradox'), treated with such suspicion in Europe, when the first specimens arrived there, that, courtesy of the eminent German anatomist Johann Blumenbach, that very doubt became part of its name.

Is that an end to the matter then? Not quite. Field, as his juggling of Nature and Creation testifies, is clearly challenged by the southern continent's presentation of creatures – the kangaroo, the echidna, the platypus – that are outside known creation and seem to rock its first premises. The challenge was widespread and the numerous attempts to dovetail creationism with such developments in natural science –

catastrophism, separate creation, the 'natural theology' of William Paley, Lamarckian evolution – are the seed-bed of a new age of human thought and will see one of these creatures, the paradoxical and apparently system-defying platypus, become a cornerstone of Darwinian theory, and eventually be dubbed 'the animal of all time'.[5]

Field, ironically, is one of the men-on-the-spot at the epicentre of this upheaval, and his 'Kangaroo' is close enough to the epicentre of his own thinking. But where, in all his juggling, he eventually fits, in the range of natural/theological options opening to him at the time, perhaps does not matter nearly as much as the mere fact that he *is thinking*. For what is it, after all, about this poem that sets it apart, if it is not the earnest and *intellectual* nature of its encounter?

It may seem strange, then, that in the whole history of Australian literature, there are few if any poems quite like it. One exception is a striking piece of the same title by D.H. Lawrence, who visited Australia in 1922 (and also wrote a novel, *Kangaroo*, that is not about the kangaroo at all). Like Field, Lawrence, in his attempt to describe his subject, resorts to hybridity: his kangaroo is part rabbit, part hare, part matron (with 'drooping Victorian shoulders'), and also, most intriguingly (but this is for another essay), part python. Were it not for Field's poem, however, Lawrence would be close to a stand-alone. Why might this be so?

One answer is perhaps obvious. Whatever else it may be that sets Field's and Lawrence's poems apart, each performs a work of *introduction*. Their implied audience is not one well acquainted with the animal they describe. Far from it. The poets may at the same time be performing an act for themselves, helping themselves understand a creature who is strange to and intrigues them (Lawrence's poem, like Field's, is an *intellectual* encounter), but they are also trying to present it to others who do not know it, by combining – using a vocabulary of – creatures they *do* know. Whether we see this, as might well be done, as an act of intellectual colonisation, or, instead, as I think it might also be seen, as an attempt, through a kind of *bricolage* (all that hybridity), to apprise a cultural Other, they are trying to bring into their own imaginary something that is outside it. Each of them is *writing*

---

5    Mervyn Griffiths, cited in Moyal (2001), xiii and *passim*.

*back* to England. An Australian audience, even in Field's time and with, comparatively, so little knowledge as yet of the country they are living in, may enjoy the description, but, having encountered the creature being described, will not *need* it in the same way. In something like a demonstration of the maxim that familiarity breeds contempt, their acquaintance will get in the way of and be far less likely to produce this kind of effort. This kind of work – this mode – represents a particular moment of cultural encounter. It's as if the kangaroo has not yet, for the percipient, or the percipient's audience, the status of thing-in-itself.

That status of thing (*creature*)-in-itself, however, has a double nature. Australians, in their familiarity with the kangaroo, might (as I've just suggested) accord it such status without *attending* to the creature at all – but there is also and always the possibility, surely, of a more attent and existential encounter. Arguably this kind of perception, the maintenance of *fresh seeing* against the deadening effects of familiarity, is one of the roles and gifts of poetry. And it is not, after all, as if the *concerns* of Field's and Lawrence's poems have disappeared; indeed the kind of epistemological anxiety – the anxiety of and about knowledge itself – that is caught, in these poems, in the abrading of the mythological and the real hybrids, so many of them part woman, part serpent (Echidna, we might note, while 'in reality' a baffling monotreme, is also the daughter of Tartarus and Gaia, part woman, part viper, the 'Mother of Monsters'), continues in Australian poetry through Christopher Brennan's (and Henry Kendall's) Lilith,[6] right through to the poetry of A.D. Hope (see the echidna in his 'The Drifting Continent') or (a Lilith again) Robert Adamson. But somehow, after Field, and if with the exception of Lawrence a hundred years later, the vein, as far as the *kangaroo* is concerned, has panned out.

What *has* happened to discount the kangaroo in the national imaginary? To the dulling effect of familiarity just cited one might add various changes in aesthetic – the pervasive postmodernist turn from landscape and the 'natural' world, for example – but they would not explain the persistence of this intellectual vector (this *epistemological*

---

6    In Jewish mythology, Adam's first wife, rebellious, carnal, more intelligent than he. When cast out by God, she is said to have gone to live with and (Echidna-like) mothered demons.

*anxiety*) in other forms, and would seem in any case to pale in the face of one major and obvious alternative.

While not a true ruminant, like sheep and cattle, the kangaroo is a pseudo-ruminant, still ideally adapted to a diet of grassy material, and so (like the rabbit) a keen competitor for the same food supply as the sheep and cattle the raising and grazing of whom – this said with no disrespect to them, who did not at any point *choose* to come – have transformed (some would say *trampled*) Australia. The clearing of land in order to supply pasture has been in some ways the kangaroo's dream-come-true, but also its nightmare. It is not the kangaroo's fault that its status as competitor has increased in almost exact step with the rise of the wool industry, but one would have to say that that has been its fate (one other thing it shares with sheep and cattle: the very thing that fattens it will see it slaughtered). For close to 190 years now the kangaroo has been seen first and foremost as pest.

There is, that is to say, another back-story, or rather a forward one. The establishment of wool in the colony. A matter involving various figures in its early stages, Henry Waterhouse, William Cox, Samuel Marsden, but principally the voracious John Macarthur, who had perceived, very early, the possibilities of breeding sheep in the colony for something other than the purpose of supplying mutton. Macarthur purchased his first flock of sheep, *to* supply mutton, in 1795. In 1797, with broader ambitions, he, along with Marsden, purchased a small portion of a tiny flock (13!) of Spanish merinos imported from the Cape. By 1801 he had the largest flock of sheep in the colony. In that year, however, he hit a snag. Still a serving officer in the New South Wales Corps, he was challenged to a duel by his commanding officer and wounded him seriously. For this and other offences he was arrested under the orders of Philip Gidley King, third governor of New South Wales, and sent to England for trial, since it was felt by King that no effective prosecution of the now rich and very influential Macarthur could be held in the colony. Although the English trial foundered, Macarthur did not return to New South Wales until 1805. His appetite for land acquisition continued to antagonise authorities, however, and within three years relations with a new governor, William Bligh (of the *Bounty*), had so deteriorated that Macarthur helped lead a rebellion against him, the Rum Rebellion, that saw Bligh flee the colony to

Hobart, where he remained, a governor in exile, until 1810, shortly before the appointment of the new, fifth governor, Lachlan Macquarie.

From 1810 to 1817, the year of Barron Field's arrival in the colony, Macarthur was in effect (and again) in exile in England, fearing arrest should he return to New South Wales. Throughout this period his wool ambitions were on hold. Upon his return, however, he pursued them vigorously. One might say that the years 1818–22 were the true cradle of the Australian industry. If in 1818 he could write to Walter Davidson 'a most pessimistic account of his attempt to introduce the merino which "still creeps on almost unheeded", observing that he sold less than ten rams a year', in 1822, as the *Australian Dictionary of Biography* tells us, 'the Society of Arts in London presented Macarthur with two gold medals, one for importing 15,000 lbs. of fine wool from New South Wales, the other for importing fine wool equal to the best Saxon; in 1824 a larger medal was awarded for importing the largest quantity of fine wool'.[7]

A small but significant irony is that if it is the wool industry that has turned the kangaroo from object of awe, wonder and intellectual inspiration to, in so much of our impoverished imaginary, pest and object of murderous hatred, it was the purported founder of that industry who hounded the author of 'The Kangaroo' from the colony. The sixth governor of New South Wales, the *Australian Dictionary of Biography* explains, 'Sir Thomas Brisbane, under whose command Macarthur's son Edward had served':

> was impressed by Macarthur and his talents, and found his opinion reinforced by 'friends' in England. Brisbane's favour revived disturbance in New South Wales in 1822 when he made known his intention to appoint Macarthur to the magistracy. This proposal produced such opposition, culminating in an official protest from Judge Advocate John Wylde and Judge Barron Field, that Brisbane had to withdraw his offer, but the reverberations of Macarthur's injured dignity and wrath reached as far as London, producing the suggestion from Bathurst [Secretary of State for

---

7    *Australian Dictionary of Biography*, http://adb.anu.edu.au/biography/macarthur-john-2390.

War and the Colonies] that the magistracy be offered to one of Macarthur's sons should either feel anxious to undertake the duties of this office. Both declined and Field was pursued by Macarthur's vituperation till he left the colony in 1824.

# Meeting Place

## Reading Derrida reading D.H. Lawrence's 'Snake'

*Beyond the shadow of the ship,*
*I watched the water-snakes:*
*They moved in tracks of shining white,*
*And when they reared, the elfish light*
*Fell off in hoary flakes.*

*Within the shadow of the ship*
*I watched their rich attire:*
*Blue, glossy green, and velvet black,*
*They coiled and swam; and every track*
*Was a flash of golden fire.*

*O happy living things! no tongue*
*Their beauty might declare:*
*A spring of love gushed from my heart,*
*And I blessed them unaware:*
*Sure my saint took pity on me,*
*And I blessed them unaware.*

*The self-same moment I could pray;*
*And from my neck so free*
*The Albatross fell off, and sank*
*Like lead into the sea.*

I WON'T SAY THAT EVERY English-language schoolchild with an interest in literature knows this passage, but a great many do – the climax of one of the great Romantic poems, Coleridge's 'The Rime of the Ancient Mariner'; the point at which, having brought a curse upon himself and his ship by thoughtlessly shooting an albatross, the mariner, all companions now dead and the albatross grafted about his neck in punishment, blesses the sea-snakes and, the albatross falling away, is at last able to pray. It may be too much to speak of this moment as a relinquishing of dominion, perhaps not even a relegation: let's just call it a break in the clouds.

D.H. Lawrence rehearses this moment in his poem 'Snake', and adds a vital twist. 'A snake came to my water-trough', he begins:

On a hot, hot day, and I in pyjamas for the heat,
To drink there.

In the deep, strange-scented shade of the great dark carob-tree
I came down the steps with my pitcher
And must wait, must stand and wait, for there he was at the
    trough before me.

He reached down from a fissure in the earth-wall in the gloom
And trailed his yellow-brown slackness soft-bellied down, over
    the edge of the stone trough
And rested his throat upon the stone bottom,
And where the water had dripped from the tap, in a small
    clearness,
He sipped with his straight mouth,
Softly drank through his straight gums, into his slack long body,
Silently.

Someone was before me at my water-trough,
And I, like a second comer, waiting.

He lifted his head from his drinking, as cattle do,
And looked at me vaguely, as drinking cattle do,
And flickered his two-forked tongue from his lips, and mused a
    moment,
And stooped and drank a little more,
Being earth-brown, earth-golden from the burning bowels of the
    earth
On the day of Sicilian July, with Etna smoking.

Reference to the Coleridge is not yet foregrounded – direct mention of the albatross is still to come – but, once cued, is obvious enough: this is not a ship at sea, but a house in a hot landscape, though (house, ship) every bit as much an image of society and its ethos; this is not a water-snake, but he has come for water; the heat is not from the sea burning 'a still and awful red' in the ship's shadow, but it is 'a hot, hot day ... of Sicilian July, with Etna smoking'; Coleridge's snakes are not golden like this one, but their 'every track / Was a flash of golden fire'; and although it may not quite be the 'spring of love' which gushes from the mariner's heart, there is, in Lawrence's persona's response, and against his cultural conditioning, an intense attraction, a desire to communicate, a sense of having been honoured by the snake's appearance:

The voice of my education said to me
He must be killed,
For in Sicily the black, black snakes are innocent, the gold are
    venomous.

And voices in me said, If you were a man
You would take a stick and break him now, and finish him off.

But must I confess how I liked him,
How glad I was he had come like a guest in quiet, to drink at my
      water-trough
And depart peaceful, pacified, and thankless,
Into the burning bowels of this earth?

Was it cowardice, that I dared not kill him?
Was it perversity, that I longed to talk to him?
Was it humility, to feel so honoured?
I felt so honoured.[1]

The voices of his past, his culture, telling him to kill, are not stilled, but he resists them. This is a point we should register carefully here. The lines don't flow as easily as they might. There's a kind of resistance in their very structure. 'But must I confess', the persona asks – and so points up the fact that there may be those who might expect him to confess such a thing, who might see it as some sort of sin or violation – 'how I liked him …?' And then 'Was it cowardice, that I dared not kill him?', and 'Was it perversity, that I longed to talk to him?', 'Was it humility, to feel so honoured?', in such a manner as isolates and asks us to examine the terms 'cowardice', 'perversity', 'humility', as if Lawrence would question their provenance, probe and possibly up-end their binaries.

And he continues watching, as if mesmerised. While the snake drinks, lifts his head, fails to acknowledge – but it is not that, for, as the poem soon implies, the very sense that such a thing might be a failure is itself a failure, so let us say 'seems oblivious to' – his watcher, and retires into his hole in the earthen wall of the persona's dwelling:

---

1    For comparability's sake I have used, as my copy text of Lawrence's poem, that presented by the editors of *The Beast & the Sovereign*, a version virtually identical to that in *D.H. Lawrence: Selected Poems*, ed. Keith Sagar (1972).

And yet those voices:
*If you were not afraid, you would kill him!*

And truly I was afraid, I was most afraid,
But even so, honoured still more
That he should seek my hospitality
From out the dark door of the secret earth.

He drank enough
And lifted his head, dreamily, as one who has drunken,
And flickered his tongue like a forked night on the air, so black,
Seeming to lick his lips,
And looked around like a god, unseeing, into the air,
And slowly turned his head,
And slowly, very slowly, as if thrice adream,
Proceeded to draw his slow length curving round
And climb again the broken bank of my wall-face.

The hole from which the snake has come and to which he returns is variously figured. At first it is 'a fissure in the earth-wall' – no indication *which* wall, though we might ask ourselves if there is any difference between saying 'earth wall' and 'earth-wall' here – and then, in three places before the next direct mention, is referred to only obliquely, although in such a manner as builds up a substantial weight of signification: the snake is 'earth-golden from the burning bowels of the earth'; he will depart 'peaceful, pacified, and thankless, / Into the burning bowels of this earth'; he has come 'From out the dark door of the secret earth'.

But then comes the second direct mention of the fissure, transformed now into 'that dreadful hole', 'that horrid black hole', somewhere, and most intriguingly, in 'the broken bank of my wall-face':

And as he put his head into that dreadful hole,
And as he slowly drew up, snake-easing his shoulders, and
    entered farther,
A sort of horror, a sort of protest against his withdrawing into
    that horrid black hole,
Deliberately going into the blackness, and slowly drawing
    himself after,
Overcame me now his back was turned.

The 'broken bank of my wall-face'. It's hard to keep the idea of bankruptcy from one's mind – bankruptcy of one's ethos, one's logos, one's culture – particularly as it is at this point that Lawrence's poem states its difference from, adds its twist to, the Coleridge. In the Coleridge the mariner is rewarded for his blessing, his hospitality (or, rather, his regret at having violated the deep law of hospitality), but here in the Lawrence, rather than finding himself rewarded, his gesture of hospitality reciprocated, his 'guest' turns and moves off, as if not noticing.

The snake is *other*, is unconcerned with – occupies a different place from – the persona's desire for recognition and reciprocality, and he is left with a feeling of failure, of having encountered something he does not understand, something to whom his own desire for confirmation means nothing. The snake – the animal; the non-human – is not *about him*, will not *stay in the light* in order to complement and complete the human, and the human, consequently, is threatened, destabilised. The picture – the face – that the animal might have completed is now seen to be broken, fissured, *holed*. The term 'wall-face' is ostended here: as if one's face – perhaps both in the sense of what one presents to the world as marking one as human, individual, and in the sense of one's *pride*, one's *dignity* (for what else happens in this poem if it is not the persona's *losing face*?) – were a barrier, a wall one puts up between one's self and a greater mode or field of being, a wall which in this poem has been breached. And that breaching, and the consequent losing of face, elicit a violent reaction:

I looked round, I put down my pitcher,
I picked up a clumsy log
And threw it at the water-trough with a clatter.

I think it did not hit him,
But suddenly that part of him that was left behind convulsed in
    undignified haste.
Writhed like lightning, and was gone
Into the black hole, the earth-lipped fissure in the wall-front,
At which, in the intense still noon, I stared with fascination.

The persona, in his anger and embarrassment, looks around, puts down his pitcher – his water-jug – and picks up a piece of wood, which he throws at the disappearing snake. He does not kill the snake, but might have; the act may not have had that specific intention, but is murderous nonetheless. Coleridge's order of events has been reversed. Lawrence's poem, in its closing, comes to a moment of transgression comparable to that with which Coleridge's starts. There is nothing elegant about this reaction. The word 'clumsy' condemns it directly. It is a failure, a shameful retreat into brutality.

It's only now the word 'human' enters the poem and we realise that, in its absence so far, there's been at least a hint – a trace – of openness, a space of possibility. But now the human and the brutal, the clumsy, are allied. Not only is the persona unable to follow the snake – how could he? this is not a matter of following – and unable to accept his (in)difference, but his inability to allow the snake to depart in peace indicates a frustration that, ineffable as it might have been, something he might have had, have shared, is being taken out of his reach. There's a sense here of thwarted possession, of a felt *right* denied.

To do him justice the persona seems aware of this, is immediately ashamed, as if he feels that other voices than his own, voices like those that had earlier urged him to kill, have for a moment overwhelmed him. The ethos that has governed him – the wall that's kept the greater being at bay – has reasserted itself. Call them what we will – of host and guest; of gift; of exchange – he is left acutely aware of the self-defeating pettiness of human rituals:

And immediately I regretted it.
I thought how paltry, how vulgar, what a mean act!
I despised myself and the voices of my accursed human education.

And I thought of the albatross
And I wished he would come back, my snake.

For he seemed to me again like a king,
Like a king in exile, uncrowned in the underworld,
Now due to be crowned again.

And so, I missed my chance with one of the lords
Of life.
And I have something to expiate:
A pettiness.

'And I thought of the albatross'. And the snake, in effect, *becomes* his albatross. He may not have killed the snake but the violence (Derrida might have said the *intellectual* violence), the retreat to the default-position of Dominion, was there. And now has to be expiated.

DERRIDA DISCUSSES Lawrence's poem in the ninth (27 February 2002) of his seminars on 'The Beast & the Sovereign', now gathered in two volumes under that title.

One can see why he might be drawn to examine it. Like Lawrence – and although neither of them name it as such – he is concerned with the species barrier, though whether on Derrida's part this is to penetrate or preserve it, whether to allow 'the Question of the Animal' (which he implies, in *The Animal That Therefore I Am*, is perhaps the greatest challenge facing Philosophy[2]) to do its work or to defuse it, is another

---

2   'I won't go back over arguments of a theoretical or philosophical kind, or in
    what we can call a deconstructive style, arguments that for a very long time,
    since I began writing in fact, I believe I have dedicated to the question of the
    living and of the living animal. For me that will always have been the most
    important and decisive question. I have addressed it a thousand times, either

matter. Lawrence's persona's encounter with the snake is not, after all, unrelated to Derrida's own bathroom encounter with his cat. And, like Lawrence, Derrida is much concerned with a 'horrid black hole', although in Derrida's case this is presented somewhat more evidently as a castration anxiety.[3]

The cat in Derrida's bathroom stares not at the philosopher as a whole, a difficult thing to do at the best of times, but at his penis: the 'horrid hole' Derrida fears – since this scene he sets up is of course calculatedly allegoric – is the hole in the Logos once the phallus at its centre (*phallogocentricity*) has been removed, as clearly Derrida feels is likely should too much quarter be ceded to 'the Animal' (from which, 'sly as a fox',[4] he seeks even to remove its name[5]).

---

directly or obliquely, by means of readings of *all* the philosophers I have taken an interest in.' (34)

3   *The Animal That Therefore I Am*, it seems to me, teeters at this point between *two* holes, the 'horrid black hole' that would be left by the (cat's-gaze-)threatened castration/removal of the phallus from the core of the logos, and the hole down which Alice follows the white rabbit, yet which, for all his play upon *je suis* as both 'I am' and 'I follow', Derrida himself does not go down. (See my discussion of *Suivre* in the extended version of 'The Loaded Cat' in Tink and Bezan, eds, *Seeing Animals*.)

Arguably, there is a hole in Derrida's text itself, a pit to the edge of which his own appalled apprehension of the immensity of animal suffering, and consequent need to *name* unequivocally, threatens to drag him, a hole that deeply unsettles him because it represents a potential undermining of so much of what he has so long argued for; insisting that he speak, and indeed *having* him briefly speak, in a manner in which he has always said we do not and cannot speak. A hole in philosophy which, in this sense, *disproves* him, or threatens to, and in so doing sets up a challenge which for the moment he will not or cannot take up. The phallus may not be removed from the Logos per se – the argument about that will go on – but it may have been plucked (or threatened to be plucked) from his own understanding of that Logos. It is perhaps for this reason that, although the cat challenges, the snake turns, and the white rabbit runs past, and although nagged by the idea that he must, Derrida cannot join them.

4   The session preceding this on Lawrence's 'Snake' is devoted to La Fontaine's 'The Wolf and the Lamb', and begins 'The phallus, I mean the *phallus*, is it proper to man?'.

5   By which I refer to Derrida's eminently defensible but also strangely self-defeating critique of the term 'animal' as an intellectual violence (see

At the outset of his discussion, as if to orient himself, Derrida focuses, via Levinas, upon a matter of *face*:

> The question is whether the snake has a head – since the question of the head has also come back regularly [in the seminars] – a head, i.e. a face and a visage, and I had recalled, I think, the question that Levinas sent back to a questioner who asked him: 'Can one say of the animal what you say of man in his ethical dimension?' – and you know that for Levinas, the other, in its ethical dimension, is what he calls a face, a 'face,' the face being not only what is seen or what sees, but also what speaks, what hears speech, and therefore it's to a face that our ethical responsibility is addressed. (237)

To the extent that this explains and substantiates the idea of *losing* face (and to say nothing of Levinas' problematic[6] insistence upon *face* in the first place; the insistence that, to be taken seriously as a sentient being, a creature must first look as much as possible like a *human* animal, is one of the cornerstones of the species barrier that I would argue Lawrence attempts to see through or across), I would concur. Derrida makes a clear point, later, that the snake *has* a face, and so is available as an ethical agent. But, while this provides us with another brief and tenuous meeting-place, it is also one of our points of departure, for, although Derrida proceeds through the poem section by section, much as I have done, he passes over – has nothing to say about – that crucial line 'And

---

footnote 1 on page 30). Were we to accept Derrida's point that, in speaking of 'the animal', we should in fact *list specifically* those animals of which we speak, we would have, to speak in effect of 'the animal' as we might think we understand it, at every point to list so many specific creatures that the vastness of the task would almost certainly reduce us to silence, and so, rather than closing the gap between human and non-human animals, as he is generally taken to be attempting to do, Derrida might in effect be only widening it.

6    Kevin Hart, for example, cautions that for Levinas '"visage" strictly means "proximity" and not anything visual. This is made clear in *Autrement qu'être*. Of course, E.L. is not always consistent and is sometimes led astray so that he presumes a certain plasticity of "le visage"' (in correspondence). This matter is discussed more extensively in chapter two (58–78) of Calarco's *Zoographies*.

climb again the broken bank of my wall-face'. Ironically – that is, for the master of deconstruction – he would seem to miss what I would argue is one of the key points of Lawrence's poem, that is, that it is *about* the undermining, the *deconstruction*, of face (he makes no mention, either, of the fact that the 'horrid black hole' into which the snake disappears – an 'earth-*lipped* fissure' – is *in* that face, indeed may be the mouth itself).

Derrida also makes much, throughout his discussion, of the ritual of hospitality – of the persona's role as host at the water-trough, of the snake's role as guest, and of the former's eventual violation of the code of hospitality. 'They go together,' he tells us, 'exile and hospitality, those asking for hospitality are exiles', and he speaks of 'the fact of being at home away from home [as] a scene of both exile and hospitality: the exiled, those asking for asylum and hospitality are not at home, they are seeking a home, and here is the man who takes them in or not, at his water-hole that is a water-source, a resource for the guests or guest-exiles or those seeking asylum' (246), but in so doing he seems to show more interest in his own theory of hospitality than in the details of the poem itself.

It is true that in Coleridge's poem the mariner's crime has been, very understandably, seen as a violation of the ritual and obligation of hospitality, and that Lawrence, as his poem closes, places his persona in a situation of similar violation, but this is not all he does. Derrida does not seem to recognise that the persona's violation of the code of hospitality – his throwing of the 'clumsy log' – occurs *because the snake has not asked* for hospitality, because the ritual of hospitality appears to mean nothing to him. Paradoxically, in 'violating' the ritual of hospitality as he does, Lawrence's persona is also asserting *and being closed in* by that ritual, in such a way as makes *him* the exile from the greater world he has just glimpsed. Read in this manner, Lawrence's poem is not an affirmation, but a critique, of the ritual of hospitality.

A part of the problem would seem to be that Derrida, as he reads Lawrence's poem, does not have the Coleridge in mind. For a Francophone undergraduate, even a graduate, student of literature this might be excusable; for a scholar of Derrida's erudition, who for decades has lived and taught literature at a high level within Anglophone cultures and at Anglophone universities, it might seem more like thoughtlessness, even wilful occlusion. And yet how else to read the complete absence of any mention by him of Coleridge's poem, and his clear *mis*-reading of the albatross?

'If we wanted to dwell longer with this poem', he writes,

> we'd spend more time looking at the other animals, such as the albatross. Why the albatross? The snake is a reptile, the low, an animal of the earth, of humus (humility, humus[7]), and that is why he [Lawrence] keeps emphasising the earth. The motif is that of the earth. So there is the low, the animal that is the lowest, the snake, and then the albatross, the animal of the heights. And as you have already noticed ... our question ... is that of the opposition between low and high. The sovereign, in principle ... is the being of height, of grandeur, of erection, his Highness. The albatross. (245)

If Derrida has any poem in mind here it is not 'The Rime of the Ancient Mariner' – if he had that in mind he'd not *need* to ask 'Why the albatross?' – but Baudelaire's 'L'albatros', a poem which, as it happens, *is* conscious of the Coleridge, and which, more to Derrida's immediate purpose, emphasises sovereignty, height and exile:

> Souvent, pour s'amuser, les hommes d'équipage
> Prennent des albatros, vastes oiseaux des mers,
> Qui suivent, indolents compagnons de voyage,
> Le navire glissant sur les gouffres amers.
>
> À peine les ont-ils déposés sur les planches,
> Que ces rois de l'azur, maladroits et honteux,
> Laissent piteusement leurs grandes ailes blanches
> Comme des avirons traîner à côté d'eux.
>
> Ce voyageur ailé, comme il est gauche et veule!
> Lui, naguère si beau, qu'il est comique et laid!

---

7   I am surprised that he does not add *human* here (humus, humility, human).

*L'un agace son bec avec un brûle-gueule,*
*L'autre mime, en boitant, l'infirme qui volait!*

*Le Poète est semblable au prince des nuées*
*Qui hante la tempête et se rit de l'archer;*
*Exilé sur le sol au milieu des huées,*
*Ses ailes de géant l'empêchent de marcher.*

– a poem a key phrase ('ces rois de l'azur') in the sixth line of which, intriguingly, given Derrida's title, Richard Howard translates as 'this sovereign of space'. I won't steal his phrase; the translation which follows is my own:

Often, to pass their time on board, sailors
will shoot an albatross, one of those huge birds
who languidly follow, like companion voyagers,
vessels gliding over the sea's abyss.

On finding themselves deposed, these kings
of sky – upon the deck, goaded by boors –
let, alongside them, their great white wings
trail piteously like unshipped oars.

How comical he is, awkward and weak,
this winged traveller only lately so assured,
one sailor shoves a pipe into his beak,
another mocks the spastic who once soared!

The Poet's like a prince of clouds
who rides the storm, laughs at their stings;
exiled on earth, amongst jeering crowds,
he's robbed of walking by his giant wings.

Perhaps distracted by his own thesis concerning sovereignty and the bestial, host and guest/exile, it seems Derrida has in effect exiled

105

himself from the wider poem, and so, ironically, placed himself in the position in which Lawrence has placed his persona.

To be fair, something in Derrida's discussion – it may not be Derrida himself – seems to know this. 'You get the scene', he writes, ironically at almost exactly the point where he might, but doesn't, say something about 'the broken bank of my wall-face':

> the snake is withdrawing, returning into its night, and the horror submerges [the narrator], 'A sort of horror ... overcame me now his back was turned.'

> I looked round, put down my pitcher,
> I picked up a clumsy log
> And threw it at the water-trough with a clatter.
> I think it did not hit him.

Which leads one to suppose that he threw his pitcher* like a projectile, an offensive weapon, at the snake's head.

Somehow – in a trice, one might say – the *log* has become the *pitcher* in Derrida's mind. The editors of the volume do not miss this, and provide, at the place I've asterisked, a note saying that 'Derrida says "cruche" [pitcher] instead of "bûche" [log]', but while this might lead us into an entertaining digression on play, slips of the tongue, and the pun in Derrida, and even, with regard to the latter (*bûche/bouche* [mouth]) lead us further into parts of the Lawrence upon which we have only touched so far (the Bocca della Verità, with *its* dark hole, *its* 'wall-face', seems to haunt this poem), there may be a different explanation.

At the centre of one of his most famous essays ('Das Ding'/ 'The Thing' [1950] 1971), Martin Heidegger, whom Derrida readily acknowledges as one of his masters – indeed it is with a mention of Heidegger that his seminar on 'Snake' closes – writes extensively of the *jug* (or *pitcher*). I offer only a taste:

> The jug's jug-character consists in the poured gift of the pouring out. Even the empty jug retains its nature by virtue of the poured

gift, even though the empty jug does not admit of a giving out. But this nonadmission belongs to the jug and to it alone. A scythe, by contrast, or a hammer is capable of a nonadmission of this giving.

The giving of the outpouring can be a drink. The outpouring gives water, it gives wine to drink.

The spring stays on in the water of the gift. In the spring the rock dwells, and in the rock dwells the dark slumber of the earth, which receives the rain and dew of the sky. In the water of the spring dwells the marriage of sky and earth. It stays in the wine given by the fruit of the vine, the fruit in which the earth's nourishment and the sky's sun are betrothed to one another. In the gift of water, in the gift of wine, sky and earth dwell. But the gift of the outpouring is what makes the jug a jug. In the jugness of the jug, sky and earth dwell. (170)

Can it be this? Why *would* it be this? Heidegger makes jug and gift, jug and hospitality, virtually synonymous. Can it be that, in an exasperation almost exquisitely close to that of Lawrence's own persona, as if he realises its impotence, its inadequacy, that extent to which it has become a prison, Derrida here unconsciously shatters – for how can we doubt that it would shatter – the very ritual he seeks to uphold?

Is it Heidegger whom, unconsciously, Derrida has Lawrence's persona hurl at the disappearing snake? Is it Philosophy?

# At Duino

## Rilke and the Killing of the Doves

I WANT TO CONSIDER the eleventh poem in the second book of Rilke's *Sonnets to Orpheus*. Here is a translation by Stephen Mitchell (2009):

Many calmly established rules of death have arisen,
ever-conquering man, since you acquired a taste
for hunting; yet more than all traps, I know you, sailcloths of linen
that used to be hung down into the caverns of Karst.

Gently they lowered you in as if you were a signal
to celebrate peace. But then: the boy began shaking your side,
– and suddenly, from the caves, the darkness would fling out a handful
of pale doves into the day ... But even that is all right.

Let every last twinge of pity be far from those who look on -
far not just from the conscience of the vigilant, steadfast hunter
who fulfils what time has allowed.

*Killing too is a form of our ancient wandering affliction ...*
When the mind stays serene, whatever
happens to us is good.

Rilke wrote this poem at the Château de Muzot in Switzerland in 1922, during the storm of creativity that enabled him to complete not only the remarkable *Sonnets*, all fifty-five of which were composed, with others, within a three-week period, but, also and primarily, the great elegies he'd begun while staying at Princess Marie von Thurn und Taxis' castle at Duino, fifteen kilometres west of Trieste on the Adriatic coast – the 'Austrian littoral' – ten years before.[1] At this time Duino, though a place famous enough in history (Dante supposedly stopped there), was little more than a fishing village, albeit at a point where the geological formation known as the Karst, a limestone plateau riddled with caverns created by underground watercourses, quite dramatically reaches the sea.

In the poem Rilke is remembering a dove-hunting expedition to a grotto in the Karst he took soon after arriving at Duino, and which he described shortly afterwards, in a letter of 31 October 1911 to Katharina Kippenberg:

> I have just been in one of the Karst dolines hunting for pigeons, quietly eating juniper berries while the hunters forgot me for the beautiful wild pigeons.[2]

Mitchell (264) cites a note of Rilke's own concerning the way in which, according to an ancient Karst custom, 'hunters … lower large pieces of cloth into the caverns and … shake them', shooting the frightened doves 'during their terrified escape'.

The grotto concerned was very likely the Pozzo dei Colombi ('Well of the Doves'), within a short walking distance (two kilometres) from Duino Castle, a possibility strengthened by the facts that, in the poem, Rilke refers to the cloth lowered into the cavern as a strip of *sailcloth* – as might come from a fishing boat – and that the Pozzo is a *sink*-hole

---

1    While Rilke was at Schloss Duino the Princess came and went, as did other
     guests. At the time Rilke began the *Elegies* he was virtually alone in the castle,
     his only company the cook.
2    *Hier unterbrochen, bin ich inzwischen in einer der Karstdolinen mit auf*
     *Taubenjagd gewesen, still Wacholderbeeren essend während die Jäger mich*
     *vergaßen über den schönen in starken Stößen aus den tiefen Felsrichtern*
     *aufschlagenden Wildtauben.* Rilke/Kippenberg, *Briefwechsel* (1954), 33.

rather than a cave per se, and so an easier geographical formation into which to *lower* such a cloth.

Sink-holes – they are called *dolina* in the region – are created when the roof of a cave or an underground watercourse collapses. Rilke might have been further drawn to the Pozzo by the fact that, at its bottom, there runs a famous underground river, the Reka,[3] which, after a thirty-four kilometre passage from Skocjan in western Slovenia, emerges, or rather *wells up*, through several fissures in the Karst rock only a few hundred metres further south, to form the short (two kilometres) but very strong River Timavo, mentioned by Virgil, Strabo, Martial and many others.

Flowing through several underground caverns on its way – one of them the *Grotta Gigante* – the Reka would very likely remind English-language readers of Coleridge's Alph, the 'sacred river' that ran 'through caverns measureless to man' and provided such a metaphor for the well-springs of poetry. It's not entirely out of the realms of possibility that Rilke may have felt he'd come to a kind of *source* here. Indeed the poem immediately before it in *The Sonnets to Orpheus* (Book II, poem x) mentions a source that emerges 'in a hundred places' (Mitchell's translation), as if – but this is beyond proving – Rilke has the Reka's emergence in mind.

The Pozzo poem is not an uncomplicated one. Several times it addresses a 'you', for example, but the identity of that addressee varies. In fact the poem has several addressees: *ever-conquering man* in the opening lines, the *sailcloths* themselves in the next, a third party (or *parties*) as recipient of the instruction in the first part of the sestet, and a fourth, in the 'our' and 'us' of the poem's last lines, who may well be an extension or broadening of the third (for surely we, as readers, are also compelled to 'look on' the scene Rilke sets before us). We might consider too that the poem is one of a sequence addressed to Orpheus, the 'god of poetry', and is dedicated, and so also addressed, to a dead girl, Wera Ouckama Knoop, who had been a childhood friend of Rilke's daughter, and who had died of tuberculosis almost three years before, at the age of nineteen. Through the entire sequence, that is to say, there is a sense that, whatever else these sonnets may be 'about',

---

3    Literally 'the *River* River' ('reka' is 'river' in Slovenian).

they also concern the subject of poetry, and, through the poet/dead girl (Orpheus/Eurydice) underlay, the nature of the poetic impulse itself.

In the Pozzo poem this concern is explicit. The ribbons of canvas, for example, come from Rilke's 'Orpheus. Eurydice. Hermes.' (1904), where (unrolling 'like a strip of cotton') they describe the pale path through the cavernous darkness along which these three ascend from the underworld, and the rite at the poem's centre is clear: the doves are a Eurydice, the shaker of the strip of sail a Hermes, the witness (Rilke, as poet/avatar) an Orpheus, and, encompassing all, the Karst cavern is an entrance to the underworld: the hand that releases the doves is perhaps that of Hades, god of that place.

And, of course, wherever the story of Orpheus and Eurydice is involved, so too is the *gaze*: the fact that Orpheus, who has been told by Hades that he can lead Eurydice back to the surface – have her again, as his living wife, in the world above (she had died when bitten by a snake) – only if he does not turn to look at her as he leads her, nevertheless at the last moment *does* turn, *does* look, and so loses her (or *sends* her) again to the underworld, to death.

In the story, this look is a glance only. Contemporary literary theory (but is the matter limited to literature, or does it have to do with the way we look upon the world itself?) has of course turned this glance into a gaze, perhaps most notably via Maurice Blanchot's famous essay 'The Gaze of Orpheus' (1955). In the shift from *glance* to *gaze*, whatever had been caught in an *action* is therefore turned into a *principle*, a *mode of regard*: '*this is what happened*, in that particular moment', this is to say, has become '*this is what we* [as poets, but also as people trying to live their lives as *authentically*, i.e. as true to the nature of those lives, as they can] *should do*'.

Just what it *is* that happened in that particular moment – the *motivation* behind Orpheus' turning – has been the focus of continual discussion. In traditional accounts the turning is a result of the intensity of Orpheus' love for Eurydice, his anxiety that she be still following him, and (for some) his fear that Hades might be deceiving him. In more contemporary accounts – Blanchot's most prominent among them – it is a deliberate choice, a change of mind, a conscious decision not to continue in the action that will return Eurydice to the world of the living, but to leave her wrapped in the arms of death.

We can be fairly confident that, *pace* his apparent position in 'Orpheus. Eurydice. Hermes.' two decades before, and consistent with the reasons for his own rejection of psychoanalysis (see note 10), it is in this latter fashion that Rilke has come to see it. The essence of *The Sonnets to Orpheus*, as he presents it – the advance they mark in his thinking – is in their attitude to death, the way they see it not as something to be avoided or to turn one's back upon, but as something to be faced consciously, before the event, for the extent to which this authenticates and intensifies one's understanding and experience of existence.

It is this that he feels he has seen and expressed at last in sonnet II.13 (only two poems after the one under discussion), which he therefore regards as the cornerstone of the sequence.[4] 'Be ahead of all parting,' he tells us there (again in Mitchell's translation):

> as though it already were
> behind you, like the winter that has just gone by.
> For among these winters there is one so endlessly winter
> that only by wintering through it will your heart survive.
>
> Be forever dead in Eurydice – more gladly arise
> into the seamless life proclaimed in your song.
> Here, in the realm of decline, among momentary days,
> be the crystal cup that shattered even as it rang.
>
> Be – and yet know the great void where all things begin,
> the infinite source of your own most intense vibration.[5]

In the existential and essentially post-Christian environment in which poets and philosophers have worked for well over a century now, this is an appealing and perhaps inevitable position. Our death is no longer

---

4   The thirteenth sonnet of part two, he writes to Katharina Kippenberg on 2 April 1922, 'includes all the others' and 'is … the most valid of all'.

5   How this differs from Wallace Stevens' famous declaration, in 'Sunday Morning', that ' death is the mother of beauty' is a matter for discussion. Perhaps we could see the one (Rilke) as an explanation of the other.

taken care of, no longer a deferrable question; we can no longer see it as an ushering into a further place; we have the opportunity not only to take responsibility, ourselves, for its conceptualisation, but to see that it is in fact death which intensifies and gives meaning to the world in which we find ourselves.

The Pozzo poem, however, goes further. When I re-encountered it after many years, and having undergone my own animal turn, I experienced what I can only describe as a moment of dismay when I came upon the poet's apparent instruction to be, as observers, as without pity for the doves as is the hunter in his steely concentration: to have no sympathy for them, no compassion, because the killing is *right*.

The German (for immediately I tried to see this as mistranslation) seemed to leave the door open to a different reading – that we, as *readers*, should feel no pity for the hunter or for the one who observes the hunt (that, implicitly, we should *disapprove*). Consultation of numerous other translations, however, while demonstrating an alarming latitude in the representation of this poem, confirmed the first reading as a kind of consensus. And it is, of course, confirmed just as much by Rilke himself, in the line *Töten ist eine Gestalt unseres wandernden Trauerns* … (which Mitchell translates – and elaborates – as 'Killing too is a form of our ancient wandering affliction').

At first, continuing my attempt to wrench this poem into something that did not so evidently condone hunting, I took the lines that follow ('When the mind stays serene, whatever / happens to us is good') to imply that the *'wandernden Trauerns'* is *not* serene, and to withdraw the apparent sanction from the killing, but no, the position that these lines take is clearly an extension of the acceptance of death that Rilke is just about to argue in poem II.13. The purity, the serenity, are, and come from, that acceptance.

There seems to be no avoiding the fact that the Pozzo poem condones killing. We are instructed to feel no compassion, no pity, for the slaughtered doves. Rilke has hitherto – in the first and eighth of the *Elegies* – shown remarkable openness towards non-human animals, indeed has argued that, while humans are restricted by the 'interpreted world' in which they live, *non*-human animals, unrestricted by such interpretation, experiencing being more directly, can be our window to the Open. This argument had seemed to many, myself included – and despite Heidegger, who adopts the notion of the Open itself, but takes it *from* non-human animals – to have changed the register when it came to thought of animals and of poetry's relation to them. But perhaps, in our enthusiasm, we had missed something.

Attractive as it might at first appear, for example, the notion that human animals live in an 'interpreted world' and that, by implication, *non*-human animals do not, begins to tremble upon closer examination. Setting aside the initial paradox, that all we know of knowing itself is *through* interpretation, the assumption that non-human animals do not themselves live in an interpreted world is just that, an assumption. The eighth elegy explains that this means that such animals have no consciousness of death (another point taken up and I think abused by Heidegger, in his statement that animals do not *die*, they merely *perish*: a formulation he also applies to victims of the death camps) and that the animal

> feels its life as boundless,
> unfathomable, and without regard
> to its own condition: pure, like its outward gaze.
> And where we see the future, it sees all time
> and itself within all time, forever healed.

This, however, tells us only what Rilke feels about his human consciousness, and nothing about non-human animals other than that he wishes them to be different, and to maintain that difference. Rilke set down these thoughts a century ago. In that time, limited as our understanding of and openness towards them still are, we have learned that many non-human animals have an awareness of time, that many have memory, that many have language, and, perhaps most pertinently, that many not only have a clear awareness of death, but have rituals built around it, and that, although the language by which they do so may be of a kind very different from anything human animals might understand as such, and their awareness of time not so obsessive as that of human animals, their world is, like that of human animals, a *bounded* and *interpreted* one, in which they live, for all we know – and surely we must allow this to them – quite aware, in their own ways, of their condition.

At the very beginning of the eighth elegy, which Rilke had completed only two weeks before writing the Pozzo poem,[6] we find an image so close to that at the centre of the latter that we must think not only that Rilke had the one in mind as he wrote the other, but saw the Karst ritual as a powerful object correlative. 'With all its eyes', the elegy reads,

> the natural world looks out
> into the Open. Only *our* eyes are turned
> backward, and surround plant, animal, child
> like traps, as they emerge into their freedom.

We could almost stop here. An explanation of such killing, of such sanction, right before us! But I think that that would be a mistake. Instead it's as if, here, through the Pozzo poem, guided by these earlier lines, we see that, all along, a concern for poetry, and for a 'right' seeing, as in seeing-of-the-world-aright, has accompanied the concern for animals themselves, in such a manner as might eventually lead to some tension between these impulses: that an interest in non-human animals *as metaphor* risks coming into conflict with a concern for their being in itself.

At best – although this would still be disturbing, and who is to say what is best and worst here? – this could be taken as a kind of poetic opportunism. How tempting, after all, it must have been, to see the bird-trapping at the Pozzo as an illustration of the statement in the elegy! And at worst, this could be taken as an extension of an idea of death, *that need not have been so extended*, to the point of condoning killing, in a manner that is tantamount to an admission that the human condition takes precedence over the non-human, a reiteration of the ancient and tragic assumption of dominion, against which Rilke has elsewhere seemed to pull. Because *humans* have come to understand death in this way, he seems to imply, this understanding can be imposed upon *non*-humans, who yet again – once more in the

---

6    The elegy was completed on 7/8 February 1922, the sonnet written on 15/17 February (Mitchell 238, 264).

long and disastrous history of the same – are instrumentalised, albeit through the velvet glove of poetry.

Perhaps Mitchell, slipping an apparently unsponsored 'too' into his translation of the *Trauerns* line, has done no more than point this out: the *other* form of our wandering sorrow, if we follow the Orpheus story, is poetry itself; poetry and killing, in this sense, are deeply allied. Poetry, the power of it, is dependent upon our sorrow, a sorrow that must have a constant and enduring source. To allow Eurydice's return to life would be to relinquish her death as a source of profound grief, whereas to return her *to* death would be to maintain this grief and all that it implies – enable it to become a *well* of our intensifying sorrow. The killing of the doves symbolises the manner in which our grief – wound – might be kept open. 'In every human artefact,' I have suggested elsewhere, 'there is, inevitably and unavoidably, the trace of slaughter.'[7] Could this, perhaps, be another point of evidence?

To be fair, and perhaps to usher in a deeper reading, Rilke may be in conflict with himself about this. There is, in this sonnet, something like a counter-voice, as if the poem were a kind of dramatic monologue. In part this is fairly open – 'But even that is all right', for example (a stricter rendition of the German would emphasise the *that*, for the original reads 'Aber auch *das* ist im Recht'), seems, with its adversative conjunction and the isolating/focalising emphasis on '*that*', to acknowledge and respond to a voice within the poet which says that *that* is *not* all right – although in some other part it takes the more covert form of a vagueness in expression, almost an *equivocality*, that could be seen as a reluctance to fully commit to what Rilke is nonetheless willing to appear to be saying, almost as if he is prepared to leave it to the reader to second-guess and to clarify.

Hoping a translation might aid me with such ambivalences, I seem, in this regard, to have stumbled, unwittingly, upon a curious window. My earliest re-readings of the poem suggested there were two passages (lines 9–11, and 13/14) crucial to a better understanding. In assembling my small library of translations I found that each translator seemed to have a slightly different sense of these lines. Some would even *add to* them in order to clarify their position. Here is a selection for the lines

---

7    'The Fallacies'.

(9–11) urging us to feel no pity (Rilke's original, for reference, reads *'Fern von dem Schauenden sei jeglicher Hauch des Bedauerns, / nicht nur vom Jäger allein'*):

> Far from the gazer remain every emotion but gladness,
> not from the hunter alone, gathering, watchful and keen
> [Leishman, 1948]

> Far from the onlooker be every breath of pity,
> Not from the hunter merely, who, proceeding, alert,
> Fulfils that which is timely
> [Herter Norton, 1942]

> Let no gasp of pity be heard from the witness
> any more than the hunter, who does as he must, alert
> and ready to act when the moment is right.
> [Young, 1987]

> May every breath of pity be far from he who sees this,
> for it is not only the hunter who gives timely and watchful effect.
> [Temple, n/d]

Notwithstanding what would seem a fine illustration of the principle of deteriorating translation, to wit that, owing to our contemporary insistence upon originality, as each subsequent translator of a famous poem feels compelled to ensure that it is sufficiently *different* from those that have gone before, translations will drift farther and farther from the text they endeavour to present, these renditions of the lines vary so much one could be forgiven for thinking Rilke's original a thing exiled, in a world apart, unable to be carried beyond the border of its own language. In this sense these versions might just provide us with an intriguing object lesson. 'And so I hold myself back,' Rilke writes in the famous opening to the First Elegy:

and swallow the call note
of my dark sobbing. Ah, whom can we ever turn to
in our need? Not angels, not humans,
and already the knowing animals are aware
that we are not really at home in
our interpreted world.

Albeit that 'interpreted' here is already a translation (again by Stephen Mitchell), it is of the most direct kind and does not misrepresent. But (and leaving aside my reservations concerning the term) can we say that these things – the world of the interpreted poem, and the interpreted world that so isolates or estranges the human, and in which we are 'not really at home' – are connected by anything more than coincidence?

'Everyday language', writes Maurice Blanchot in 'Literature and the Right to Death' (1947), in one of the earliest statements of what has become almost a mantra of contemporary theory – that to draw something into language is in effect to kill it – 'calls a cat a cat, as if the living cat and its name were identical', yet '[t]o name the cat is ... to make it into a non-cat, a cat that has ceased to exist, has ceased to be a living cat'.

Could this or something like it be in Rilke's mind when he rehearses the Orphic rite in the Pozzo poem? Poet and hunter are there aligned: in a key line, yes, but also by the fact that the poet watches the hunter so attently that his gaze is likened to the hunter's concentration, as if the art of hunting and the art of poetry were somehow deeply akin. And then, of course, the 'facts' that the poet subsequently *makes a poem* from the incident, and that in that poem there is a line – *Töten ist eine Gestalt unseres wandernden Trauerns* – which argues poetry and hunting to be forms of the same wandering grief. The poems, or rather poetic *subjects* (doves) stirred by the poet's descent into the subconscious, cannot live past or through their becoming-poems, their emergence into light. The act of making the poem is a kind of killing. It may be claiming too much to suggest that the ghost statement behind the lines which argue that the hunter and the poet are to feel no pity – that statement which seems to tell *us* to feel no pity *for* them – is now heard to whisper a little more loudly, but it is perhaps not claiming too much to say that the poem, and Rilke's presence in the poem, and the various counter-voices we have identified, amount to a confession of sorts, an admission of complicity.

Is this it, then? That wrestling, as the poet must – as could be said to be the essence of the poet's work – to capture what has come up or been stirred from the subconscious before it inevitably goes back down, *necessitates* the gaze (or glance), and that the work itself, that putting the what-is-then-glanced into words, is also the 'killing' of that thing, the separating of itself from itself, so that this, the frighting of the doves, the hunting, becomes an allegory of poetic creation?

No, perhaps not. Perhaps that is too rigid an idealisation. Blanchot might go that far, but I don't think Rilke does. I suspect, however, that we are getting closer. The *well*, *death*, these words keep coming up, as if they – and the word *wound*; I keep waiting for *wound* (it is not, after all, as if Rilke has not himself used it[8]), as well, as constant, vital supply – mark out a key *topos* in modern poetry. I think of Lorca's famous statement of the *duende*, a term he borrowed from Spanish folklore in the attempt to explain the deep, burning, passionate power of art, and of the greatest poetry in particular:

> And the duende? The duende does not come at all unless he sees that death is possible. The duende must know beforehand that he can serenade death's house and rock those branches we all wear, branches that do not have, will never have, any consolation.
>
> With idea, sound, or gesture, the duende enjoys fighting the creator on the very rim of the well. Angel and muse escape with violin, meter, and compass; the duende wounds. In the healing of that wound, which never closes, lie the strange, invented qualities of a man's work.
>
> The magical property of a poem is to remain possessed by duende that can baptize in dark water all who look at it, for with

---

8    Mitchell quotes a letter to Witold Hulewicz of 13 November 1925, in which Rilke refers to the figure of Wera Knoop, in the *Sonnets*, as belonging 'to those powers which keep the one half of life fresh, and open toward the other, wound-open half' (250).

duende it is easier to love and understand, and one can be sure of being loved and understood. In poetry this struggle for expression and communication is sometimes fatal.[9]

In his months at Duino in 1911/12, Rilke, still to emerge from a long period of depression following the completion of *The Notebooks of Malte Laurids Brigge*, was struggling with the question as to whether to undergo psychoanalysis and corresponding extensively about it with Lou Andreas-Salomé, Baron Emil von Gebsattel and others. In the long run, in a position not uncommon for artists and writers, he decided against it, for fear that dragging his inner demons to light might reduce his creativity, as if the inner wound and one's perpetual wrestling with it were in fact a source of creative strength.[10]

This question had been with Rilke a long time. There is no Hermes figure in the original Orpheus story, for example, but Rilke adds him in 'Orpheus. Eurydice. Hermes.', to lead the death-blind Eurydice up from the underworld, almost as if he were an analyst coaxing an inner demon from an analysand. In the Pozzo poem this part would seem to be taken by the person who, shaking the sailcloth that we have seen already stands for the path upward, stirs the doves into flight (these are not inner demons per se – Rilke knows well enough that the offerings cannot be taken for the problem itself, nor would he want them to be).

The gaze, we might point out here, does two crucial things: it enables the artist or poet to apprehend that which he or she can turn into the poem or work of art, but it also sends that thing back down. The wound must be maintained. The most powerful poetry *comes from* that wound and, without that wound, it is feared that poetry will either not arrive or that, when it does, will not have the force it might have had had the wound still been a factor. As Lorca says (and Hélène Cixous most memorably after him[11]) the wound must be constantly healed

---

9    From a lecture, 'Juego y teoria del duende' ('Play and Theory of the Duende'), delivered in 1933. See Maurer (1998), 58.

10   It would seem that Freud and Rilke met only twice, and then only briefly. Legend has it that on one of these occasions, Rilke, rejecting the offer of psychoanalysis, said 'If you take away my devils, my angels may leave too'.

11   'All literature is scarry. It celebrates the wound and repeats the lesion', 'Preface: On Stigmatexts' (1998), xi.

(tended), yet constantly open. The sailcloth is lowered 'as if … a signal to celebrate peace', and in return a handful of doves – symbols of peace; the sign sent by God to indicate to Noah that he *and the (other) animals were safe* – is thrown up, only to be shot, their flight stopped, as if no peace was ever intended.

What can we take from this? That in this poem Rilke seems almost deliberately – almost against his better judgement – to have chosen a scene and subject that will trouble the surface of his poetry? That this poem seems almost to *be* a wound? That in this poem we see something we cannot be sure that Rilke wants us to see? That in this poem it is almost as if the wound – this deeper wound – looked at us?

Myths are said to encode and perpetuate – give a kind of vital and perdurable form to – the truths of the tribe. And perhaps also, in their sacralising aspect, to keep things vague. Even their perdurability may be deceptive. That some of them seem always with us – changing their clothes, as it were, but not their inner form – may, yes, indicate the acuity with which they capture something fundamental to human experience, but may also suggest we have a more obscure and ambivalent relation to them than we think. In a sequence dedicated to Orpheus himself, the Pozzo poem is arguably amongst Rilke's most Orphic. But it is Orphic in a particular manner, and perhaps we should consider that manner more carefully.

The manner in which Rilke presents the Orpheus myth may be 'true' enough in some ways – serviceable, in any case, and powerful, in the sense that it captures some of the deep structures-of-relation of poetry and indeed creativity more generally: the way a consciousness of death, of temporality, of transience, whether of others or one's own, can give to the things one surveys a mysterious, intensifying penumbra, as twilight, say, or a time just after rain can seem to intensify the colour and forms of objects; the way grief or some other forms of powerful emotion can heighten one's awareness, intensify one's feelings for and desire to keep hold of those things and beings immediately around one (near-death experiences, for example, are repeatedly said to increase one's appreciation of life). But might it not be that in these applications and interpretations there is a drift, a shift of attention, that reflects something more contemporary? Why is it that we, as poets, philosophers, literary theorists, don't think so immediately of what are doubtless some of the myth's earlier interpretations and decodings – as a register, say, of the way loving too much can be itself destructive, the way loss attends love as if they were somehow connected like the different ends of an hourglass?

What, in particular – to turn back to Rilke, for it is Rilke who, more than anyone else, has re-centralised the Orpheus myth in the modern poetic consciousness – is it that could *extend* this myth to non-human animals in the first place, and, beyond this, to the *killing* of such animals? Or, rather, to refine this question a little, what is it that could make of death so overweening a creative principle that it could so readily – as if *naturally* – *be* extended?

One suspects that a default-position has come into play here – that death has become a kind of blanket term to cover a more nuanced array of what we might call generative emotions. I will not say that that blanketing in itself is not part of an old wound; indeed that is the point, and that the misty power of myth, *this* myth, perhaps itself needs to be placed under the spectroscope. Neither would I say that Rilke does not have his own deep and generative wounds (the death of his sister, Maria, a year before his birth not only giving him his name – René Maria [Maria re-born] – but leading his mother to dress him in girl's clothing for the first six years of his life; his traumatic experiences in the military school to which his father then sent him, etc.), nor that they don't have their own very clear connections with death (one could say that Rilke had always a dead woman inside him!), but one cannot readily imagine what they might have to do with killing. The very connection itself seems a wound. It would appear that something else, a third force, has come into play, to take this death (death of the sister/death of Eurydice) and not only turn it into a governing principle of poetic inspiration, but extend it to *killing*.

In the aforementioned existential and essentially post-Christian environment – the decline/absence of a religious faith that would see it instead as permeable, a transition – death has taken on a formidable *monolithicity*. Blanchot, Bataille, Rilke, Heidegger (especially) seem fixated upon it, to the extent of a kind of *necrologocentrism*. While Rilke himself has been generative (Heidegger not only [mis-]takes his sense of the Open from him, but takes up, indeed centres, his call for a before-knowing, a before-accepting of death[12]), and there can be no doubt that history has played an enormous part, in the way in which the *Shoah*, and the First World War that paved the way to it (and in which Rilke was briefly forced to serve), have created a great crater in our moral comprehension, traumatising and focalising the philosophy, art and politics of a century, one must still ask whether this necrologos itself may be a subconscious falling-upon a default position? Is there no way of refining our understanding of it (this fore-shadow of death), or at the very least of placing it in inverted commas, under erasure, so that it seems less inviolable, less like an end-beyond-thinking, so that we might feel more able to look through, under or past it, to something else that the insistence upon death per se might be preventing us from seeing? Is it, say, a *guilt* that drives us there, that *creates* this place, this necrologos – an *excuse* wound,[13] to which, like scar tissue on a deeper wound, we thus default?

The very trembling and inconsistency in Rilke's poem seem to betoken a sense that a connection is missing, that something is not right, a hope that words, whether as assertions ('Aber auch *das* ist im Recht', '*Töten ise eine Gestalt unseres wandernden Trauerns ...*', 'Rein ist im heiteren Geist, / was an uns selber geschieht'), or through careful inspecificity, might *paper over* what is, after all, a kind of wound within the poem itself.

Would it be too much to say that this guarded space, this hidden, deeper wound – the wound we may be attempting to mask by *excuse* wounds – is our rift with non-human animals? A repressed consciousness, let's say, of the massive slaughter of which we are daily a part (something, after all, of which Rilke, as a 'vegetarian', might have had a keen awareness); a sense – bearing in mind that to sacralise is to short-circuit discussion and examination, to *place out of reach* – that the sacralising of death itself, the centralising of it, the attribution to it

---

12  'Is Rainer Maria Rilke a poet in a destitute time?', Heidegger asks in 'What Are Poets For?' (1971): 'How deeply does [his poetry] reach into the abyss? Where does the poet get to, assuming he goes where he can go? ... The time remains destitute not only because God is dead, but because mortals are hardly aware and capable even of their own mortality. Mortals have not yet come into ownership of their own nature. Death withdraws into the enigmatic. The mystery of pain remains veiled.' (94)

13  See 'The Wound' below.

of powerful generative force, enables us to give to this slaughter itself a necessary and even sacred cast, if only by that roundabout way by which hunting, by involving *ourselves* in killing, enables us to reaffirm, to reinvigorate, to inject it again with this putative sacredness, and so help us to keep our mundane, daily, nagging guilt at bay, through a kind of rite of repression, a rite of abjection.

Could it be that one wound – our maintenance of it (the monolithicity surely a cornerstone in this strategy) – is being used to repress our consciousness of another, to throw up, as it were, a handful of doves to distract us? The position, after all, is no longer as uncommon or unthinkable as it might once have been. Derrida, for example (2008, 2009/11), eventually places it at the heart of the logos.

How can we offer peace to animals, while the shield-wound is our priority, and this deeper wound is unaddressed? More specifically, how can poetry offer or work towards this peace? For my part I can't see why poetry can't continue, in the face and service of the non-human animal, to do most powerful work, but one suspects that there are some alterations to be made in its own deep image of itself, some ancient and powerful bonds to be broken or reconfigured, before its powers are freed for this service. Our treasured bond with Orpheus may be one of them.

# Foal's Bread

*Hippomanes*: small (up to 1.5 inch thick and 8 inches in diameter), circular, flat smooth proteinaceous bodies found in the allantoic fluids especially in mares, cows and camelids. Called foal's bread or foal's tongue. On cut section they are semisolid, homogenous, amber to dark brown-coloured.

    – Virginia P. Studdert et al, *Saunders Comprehensive Veterinary Dictionary*[1]

*Hippomanes*: ... ἵππο-*s* horse + μαυ-, root of μαίνεσθαι to be mad ... 'A small black fleshy substance said to occur on the forehead of a new-born foal'.

    **1601** HOLLAND *Pliny* I. 222 These foles verily, by report, haue growing on their forehead ... a little black thing of the bignesse of a fig, called *Hippomanes*

    – *Oxford English Dictionary*

ALBEIT WITH SOME TREPIDATION, I'd been looking forward to reading Gillian Mears' *Foal's Bread*, hoping for a novel that might mark a step, in what so far has been essentially a carnist national literature, towards the non-human animal. Certainly the book's title and cover – the latter

---

1    4th edition (Elsevier Health Sciences, 2011).

bearing, in the first paperback edition, a photograph of a horse's head (a foal's?) straining forward, though whether in effort or anticipation of something (an offering? a caress?) isn't clear – had led me to anticipate some such thing.

The book had received high praise. 'A blessing', read one endorsement, 'from one of Australia's finest writers.' 'Glorious', said another, '[a] bold and brilliant song of praise.' 'It is a rare fiction,' wrote a leading reviewer, 'that lavishes such attention and care on the depiction of animals.' The work received several major awards and was shortlisted for numerous others – a golden run – and overall, my few misgivings aside, I think it deserved it. At times it had me thinking of *Tess of the d'Urbervilles* or *The Mill on the Floss*, and still does.

Why, then, do I *have* misgivings?

The book opens – its first paragraph – with Cecil Childs and his only daughter, Noah, arriving at One Tree, a farm that, with Noah's marriage to Roley Nancarrow, the son of that farm, will soon become its principal locus:

> this was what they'd been doing for close on two weeks for the buyer in Sydney. Picking up pigs from one farm then the next. Cecil knew that they were a day early for loading onto the boat at Wirri that would take the pigs down to Port Lake and then to the bacon factory in Sydney, and this knowledge had also brightened his mood. (5)

The book, that is, begins with its principal character involved in droving animals to slaughter. Unsettling, to a person concerned with the elimination or at the very least reduction of such animal suffering, but in truth there's nothing remarkable or unusual in such a beginning. *Foal's Bread* is set in rural New South Wales from the mid-1920s to the 1950s, decades (we might think) before animal liberation was on anyone's mind.[2] Much of its action takes place on a dairy farm and

---

2   Of course, although the term itself may not have occurred to them, the liberation of *non*-human animals from suffering at the hands of humans has been much on the minds of many people from time immemorial. Peter Singer may have popularised the term in his Animal Liberation (1975), and

a central narrative drive has to do with the show jumping of horses. Surely the businesslike treatment of non-human animals in the work is only natural, a matter of verisimilitude on the novelist's part, is true to its time and place. Undoubtedly the attitudes of the people we encounter in the narrative are attitudes we'd have encountered in the vast bulk of the populace, rural and urban, at that time. And, whatever else they may be, the 'true' writer, the 'great' writer is also a lens, a sort of verbal camera, above it all, paring his/her fingernails, as one of them famously put it. Naturalism. Realism. Hand-in-hand. Almost inseparable, as they tend to be. *Naturo*-realism, we might call it.

Casting about at the outset for some kind of moral gyroscope – 'normal' as it might be, not all that many novels *do* start out with a herding-to-slaughter – I found myself reading the next pages with particular attention. Noah gets on her pony and, her father suggesting she might do a little jumping practice, rides up to the homestead, taking on a few fences as she goes, but feels uncomfortable, as if something within her – the baby, her uncle's – is not right. 'Slipping off' as she arrives – for a second this might almost be one of the moments of moving human/non-human animal intimacy for which the novel has been praised – 'and parting the pony's mane, she leant close to take a little taste of the salty neck'.

This, however, is when the pony,

thirsting for a drink, gave her a butt in her belly so rough that she punched the horse's nose straight back before looking up to greet the two women who were coming out of the house. (7)

---

the first Animal Liberation group so called may have been founded in Australia shortly thereafter, but veganism and animal rights had been publicly bruited concerns in Australia for a century. See, for example, Robert Jones (1888) in Crook (2014). Humans, Jones urges, should '[c]ease their consumption of that grossest of all foods, dead flesh ... The suffering of gentle, domestic animals by land and sea, in railway trucks and cattle-steamers, from thirst, hunger, cold, heat, overcrowding, fatigue, blows, terror, and sickness, not to mention their death-agonies, and the unspeakable horrors of the slaughterhouses are such as no pen can describe ...' (129).

"'Here. Here!" shouted a woman', and this reader did a split-second double-take, a mixture of alarm and something like relief – here was a woman about to say 'Stop punching that horse!' – although soon enough realised that within or behind this 'Here. Here!' there might just as likely have been a 'Hear. Hear!', in support rather than admonishment. And ultimately – my point – I am not sure that the book really helps us decide, here or at a number of points elsewhere. Rather, *like* this passage, it turns away, distracts itself. *If* there is admonishment here for a moment of cruelty towards the horse, it's fleeting, a glimpse only, unconfirmed. "'Here. Here!" shouted a woman', the full sentence reads, 'at a pair of over-eager dogs', which is to say that this comment wasn't about the pony at all. It seems like a deliberate construct, and I cannot rule out that it isn't (silence, exile, cunning). One wonders if there's a name for such a trope. It would be useful if there were, since it occurs again.

We are now, in any case, ourselves almost immediately distracted. The girl – it is hours later – gives birth, alone in the darkness, surrounded by hungry, threatening pigs, and sacrifices – but *is* it sacrifice? does that make too much of it? – the child to the waters of Flaggy Creek. Sacrifice? or infanticide. Placing him in a butter-box. We might think of Moses in the bulrushes.

In the chapters which follow, Noah meets and marries Roley, the 1926 national show jumping champion, and the couple embark upon what they hope will be a golden future. But if one hopes, in one's turn, for some clear reorientation towards non-human animals, some critique of or rapprochement in the ancient master-slave relation, one is disappointed, if only at being left uncertain.

Perhaps it's all a matter of perspective. To someone watching closely the fate of its non-human characters, the narrative, as it unrolls, becomes a record of animal abuse, accepted and relatively unquestioned (though one surely must question that unquestionedness).

We are, generally speaking, inured towards cruelty towards non-human animals. We have almost to train ourselves to see. Here and there one of Mears' characters raises an eyebrow at another's mistreatment of a non-human animal, but is as likely as not to abuse their own animal(s) when it suits them. That several of these individuals are characters central to and exonerated by the narrative tends to suggest either that this abuse is accepted by the author (harder to believe), that it's a matter of a kind of cultural blind-spot, or that it's presented, as aforementioned, in the interests of naturo-realism, historical and cultural verisimilitude. People *do* treat animals in this way, it could be seen to be telling us, albeit implicitly. It stinks, but they do. And, of course, when one considers the book's chosen style, its roots in demotic Darwinism ('survival of the fittest'), there *is* a venerable link between naturo-realism and violence, cruelty, and the normally/ nominally abject, as if these things – violence, cruelty, the physically *explicit* – were a kind of *guarantee* of the 'natural', the 'real' in the first place.

It may be, of course, that the author, within this naturo-realistic envelope, has animal cruelty consistently and deliberately within her sights, in the hope that readers will interrogate this aspect of her narrative and the society it depicts, but, if so, (a) she is perhaps too subtle, and (b) her medium – Naturalism – undermines her, in a manner I will come to.

First, however – a different kind of hermeneutic tack, every bit as naturalised – there is the matter of Noah's name. Noah. He who took the animals, two-by-two, aboard the ark, to save them from the flood. Although here, of course, Noah is a woman. When we look – Biblically, 'historically' – for a *female* Noah, we find her (at Numbers 27, 1–11) to be one of the five daughters of Zelophehad, who, in the late stages of their trek through the wilderness, petitioned Moses and the Israelite community for the right to inherit their father's property since he had died ('in his own sin') without a male heir.

Zelophehad means 'dark shadow'. The daughters were granted their request so long as, in marrying, they kept the property within their broader tribe, as each did by espousing the son of an uncle. On the one hand, then, we have Noah, the Ark, and the concern to rescue animals, and, on the other, the 'dark shadow', tribes in the wilderness, the role of the uncles, the figure of Moses. Are we to take these further into the novel – one of them? both? Although I wouldn't deny it's possible to see them, beneath the novel's surface, in a consistent and complex relation, such things, lending a kind of mythic force to the novel's proceedings, might ultimately only distract us.

'The day after Cecil Childs found himself eliminated,' begins the first of five passages selected to illustrate the abuse I've been speaking of (there are many more[3]),

> with the darkness of a hangover filling his mouth, he'd taken Rainbird down to a paddock behind the showground. On the way he grabbed a pitchfork not his own and, just starting off with the flat of it, began to hit the horse. By the time Lance Oldfield arrived, shouting a warning for him to leave off and never go near the horse again, it was too late. Cecil Childs had wounded the horse. The fork blades had gone into the rump. Also nearside, a shameful sight, all its ribs running bright with blood. (38)

'Noah took the lead rope,' begins the next passage,

> and, letting go the ear, began kicking the horse in the guts. For a moment Lainey [Noah and Roley's daughter] looked somewhere else. The thumps and thuds landing on the horse sounded so violent, like all the heads in hell banging together when the fires had gone out. (170)

The other three:

> When George [Noah and Roley's handicapped son] had finally learnt to tie a bow last year, he'd kept putting one of Lainey's green hair ribbons around a new black kitten's neck. Although she and Laine had sworn to find all of the kittens homes, it didn't stop Minna doing what she always did. Drowning even that one. With the ribbon still around its morsel of a throat Noah knew, because she'd accidentally found the sack when checking down at the creek on the irrigation pipes. (259)

---

3    Others, for example, can be found on pages 48 ('"Your father's got a point"'), 191 ('"Wire up the top railing?..."'), 194 ('In each hand he held a stick …'), 240 ('Talking to the horse again …'), 266/7 ('Without a bit of protest …'), 271 ('Before Lainey knew it …'), 277 ('For all of the morning she'd used the spurs …'), and 292 ('then she twitched …').

They found the elderly cows not far from the yards, but as if knowing this was no friendly morning, they all charged off. Then Roley's dog, turning them, got a hold of one of the old girls' noses and wouldn't let go. So that even as Lainey used a little yard whip to get that most cunning old girl into the yards, and to make the dog let go, the blood was already dripping.

'Don't let that one out,' Noah instructed Lainey, who was working the gate. 'Oh, these bloody gins and bitches. Don't know why we bother. Fair dinkum I don't.' (297)

In her fingers Noah felt what could be a nut. She knew what she had to do. Holding out her hand for the emasculators she slid them along the arm that was so far into the cavity of the cow she was up to her shoulder.

'There. I think I've got it, so now I pull it my way a bit, then just one snip. Yep, that's one. Very good.' She flung it to the waiting dog. 'I've known blokes who'd bring out lump after lump of fat … now for t'other side.'

And again, in a few moments Lainey watched the second ovary fly through the air. Grass the colour of an old biscuit, she thought, with jam trickling down.

'Oh what kind of stupid bitch am I?' Noah, out of the cow, wiped her eyes with her forearms and tipped the sack upside down.

'What's not here?'

'Phew,' said Noah, locating after all the long sack needle.

The cow, jumping at the needle threaded with bag twine going in and out, thought better of it when Noah took her ear and began to twist it up. (299)

Although her father, Cecil Childs, is cause of some of the worst of them, almost all of the other instances of abuse in the novel are at the hands of Noah. Their frequency increases as the narrative's pressure upon her intensifies, and a strong case could made for them as a calculated index of her psychological damage, whether from what at *this* point in time would be called abuse at the hands of her beloved (for for a good proportion of the novel he *is* presented as loved and much missed by her) Uncle Nip, from other aspects of her upbringing, from the early death and earlier impotence of her husband, or from her predisposition towards alcohol abuse that is itself most likely consequent upon some or all of these others just mentioned.

In some ways truly and traditionally *naturalistic* in this regard (its eye firmly on the combined imperatives of *nature* and *nurture*), *Foal's Bread* would make an intriguing psychological study. Noah is also, as intimated to us very early ('dark shadow'), and as becomes increasingly a consideration as the book develops, part Aboriginal. There is a demonstrable cultural impression, long offered as a truism (see *Coonardoo*, see *Capricornia*), that Indigenous Australians are particularly good with horses, though I'm not aware that this is very often presented hand-in-hand with violence towards them, and it may also be that we are given this particular nexus – these instances of equine abuse – as a symptom of the intensity of Noah's inner conflict.

From one perspective, it could be argued, the instances of abuse of non-human animals in this novel are both fairly flagged as such, and given a clear rationale. But does the ethical dilemma they present begin therefore to resolve itself, or does it only deepen? Noah eventually dies under a horse from which, against the better judgement of horse and rider alike, she has asked the almost impossible, that is, that the horse leap – and *carry* her – eight feet to clear a ruined bridge. A kind of suicide (she has just castrated her elderly uncle – another of the novel's expiations – for an attempted assault on her daughter, and left him bleeding to death), it could also be seen as a kind of sacrifice, the horse abuse of which she is principal agent not standing out *against* a peaceful and non-abusive society, but as a kind of scapegoat, absorbing and concentrating that society's sins and bearing them away, into a murky, abjective matrix of Aboriginality, pederasty, animal abuse, and infanticide.

For a start, can we call this abuse at all? Or, rather, for there can be no question that we should, shouldn't we also, *as* we do so, be aware of the dangers of doing so without careful qualification? I don't mean to doubt that that, abuse, is what these instances of the mistreatment of non-human animals in the book *are*. But the question is arguably more complex than this – might even, as we work on it, lead us towards a better intuitive understanding of the trope, or aporetic space, I spoke of earlier ('Here. Here.').

There's a distinct sense that the *mis*treatments to which we find ourselves witness are considered to be so only or largely out of their *excessive* nature, that is, are instances of *going too far*, and to isolate such incidents, and to explain them as products of deviant and damaged character – as much in this book, much in the nature of ' naturalism', and much in our conditioning as readers would lead us to do – is also, ironically, to *normalise* the practices that they exceed or deviate *from*. Aren't several of the passages quoted – or so we might have found ourselves thinking – simply descriptions of 'normal' procedures of animal husbandry? To call such instances of abuse excessive – which they unquestionably are – is also in effect to obscure the fact that they are part of a continuum, and the extent to which *all*, on that continuum, is abuse. To designate *excess* is to normalise that which one holds it to be an excess *of*.

There is also the manner in which such explainings-away as psychological aberration – consequences of the abuses inflicted upon human characters – take the pressure off and divert attention from the acts, and in this case the non-human victims, themselves. It perhaps takes this too far to say that in a society/situation where everyone is victim, no one is victim, and that the buck does not stop, but there is nonetheless a part-truth in it, and this part has to be considered.

Our conditioning, as readers, leads us to look for explanations of aberrant behaviour, and to look largely along familiar pathways. To what extent could this training, this conditioned hermeneutic, actually function to diffuse – to process away – our objection to such behaviour? What might, and perhaps should, be outrage, evaporates, attenuates, dilutes in the face of (a) our 'understanding', and (b) a set of tacit taboos – versions of what some would call political correctness – concerning (say) criticism of Aboriginals, criticism of victims of sexual abuse, criticism of disadvantaged women or, turning towards Roley Nancarrow (and George, his son by Noah), criticism of those with disabilities.

Don't get me wrong (though I'm imagining some readers will): I have no truck with racism, with paedophilia, with sexism, and, as someone 'disabled' by the same condition that disabled Mears, I find myself having continually to face and absorb ableism. What I *am* saying here is that *nothing* should excuse the abuse of other animals and the perpetuation of their suffering, and that the *interposition* of such proscription laden signifiers – by which I mean areas where the reader and critic are accustomed to going only on tip-toe – between the main body of the narrative and the abuse we find within it becomes a (further) means if not of masking such abuse, then of distancing it and absorbing (muting, cushioning) its moral impact.

Gillian Mears suffered from multiple sclerosis. She finished this novel at great effort and with great difficulty. She is, in this regard, to be applauded and deeply sympathised with. My argument may seem as if I am myself in some way abusing a disabled person but I think that would be the most simplistic of responses. Decades of training have given me a clear ability to know 'a fine novel' when I see it, and *Foal's Bread*, as I have said already, is just that, 'a fine novel', both masterful and powerful in so many aspects of its narrative – and with, as it happens, in the fate of Roley and George Nancarrow, a couple of Australian literature's more significant and insightful portraits of disability.

Just as clearly, however, I find that I have come to a point where 'a fine novel' is no longer enough to hold me, is in fact harder and harder to read if, in its process, it does not engage critically with society's widespread acceptance and/or wilful occlusion of animal suffering, or at least attempts, as its narrative progresses, to eschew those attitudes and actions – often, for a novel, this means those *depictions* – which support the perpetuation of such suffering. And – as evidenced by the paucity of any such works – that is no easy task. Without constant vigilance – and this is a vigilance that few even of writer-activists have the determination to maintain – writing will fall all too readily back on the default-position of a carnist and abusive society which has so normalised the instrumentalising of non-human animals that, although it is a part of the very fabric of our being, it is for the most part invisible and absent from thought. We 'abolished' slavery, but slavery continues extensively, is arguably sustained, even demanded by our lifestyle; we made vast inroads into sexism but it is every day in our faces; we deplore racism but our culture is rife with it; we deplore the suffering of non-human animals, yet every day contribute to it, on multiple levels of our being.

It's a hard road. One has to learn to *unread*, to read, as it were, against some of one's deepest conditioning as a reader – to be, from a trans-species perspective, what Judith Fetterley (1978) famously termed, from a *feminist* perspective, a *resistant* reader. In a civilisation that for thousands of years was (and arguably still is) a patriarchy, Fetterley argued, the naturalistic perspective was (and arguably still is), in a vast number of subtle and unsubtle ways, riddled with that patriarchy, presenting as 'natural' what was (/is) in fact a deeply *gendered* world. If one followed only the fate (/fight) of one's heroine, without at the same time trying to unpick the mesh in which she is so deeply entangled, one risked inheriting and/or perpetuating, rather than escaping, her prison, just as she, however much she might feel she escapes it, also takes her prison with her.

From a non-human animal(ist) perspective, 'naturalism' must be treated with the same suspicion. Carnism (the belief that one has the right to slaughter and consume the flesh of other creatures[4]) and the cruel exploitation of animals have been for so long the norm, the 'natural', that the world which an unresisting naturalism re-presents to us, both as a style of writing and a style of reading, is riddled by their filamentous roots.

A further point is harder, requires, in a sense, that the reader resists the very part of themselves that might be most activated by the question. The very *attempt to explain* characters or events in the text, as I have been doing for so much of this essay so far, might be part of the problem, since such explanation is a drawing of such things under the umbrella of what *is*, and so in itself a reinstitution of the human, a *naturalisation*, an extension of the (understood) natural. We must, consequentially, contemplate the possibility of a *resistance to explanation*. At least initially, it being arguable that naturalism has created, over our thinking, a kind of crust or scar tissue that must be cut, or broken through, so that there might occur a kind of thinking freer to perceive what the veil of naturalism currently hides.

---

4    The term *carnism* is generally accredited to Melanie Joy (2009).

Perhaps that sounds a little like Bertholt Brecht. The A-effect. Brecht argued for disruptions of naturalistic surfaces in order to keep the minds of his audience *open to thought* and so to the perception that a world they had thought inflexible, unchangeable (*because* 'natural'), was in fact a human construct and therefore negotiable. A Marxist perspective, in his case, just as in Fetterley's case it was a feminist one. I am arguing for something similar (an 'animalist' perspective?), yet also for a bifurcation, a *resistance* on the one hand, and a *vigilance* on the other, for if one essential strategy is to be a resisting reader, another is to be a *per*sisting one – to follow overlooked or unsuspected threads, along the lines suggested by Deleuze and Guattari in their account of the rhizome, to see where such threads can lead us and what internal economies they form, or to take *details* and, rather than slide over their surface, as virtually any 'naturalistic' author would have us do, to *worm in* (a positive, not a negative use of creature image: the worm inhabits the same territory as the rhizome, tills the soil in which the rhizome grows).

As we might, for example – to return this essay to its beginning – 'interrogate' the *butter-box* in which Noah's first child is set afloat in the early pages of the novel. A box that has held butter. Butter that has come from the milk of cows. Cows that have produced the milk to feed calves (read also foals?) that have been taken from them. Cows who will be made pregnant again and again, and their calves taken again and again, to keep their milk-flow going. A vector – a *rhizome* – that might lead us on to the *sterilisation of the elderly cows*, later in the book (297–99, quoted at length above).

*Why* sterilise elderly cows?

'The major welfare benefit of spaying', according to a discussion paper prepared by the Cattle Standards and Guidelines Writing Group in February 2013, 'is the prevention of unwanted mating and pregnancies thus promoting survival of female cattle not suitable for breeding due to poor conformation or not suitable for breeding and subsequent lactation due to an advanced age'[5]– for which latter option we could perhaps read '*multiparous* cows [cows that have already had several calves] *no longer* suitable for breeding ... due to advanced age'.

---

5    https://bit.ly/2Ryekkl.

And very well: this would seem to give us an ostensible reason for Noah's sterilising them in the scene earlier detailed. But of course it is not that, even ostensibly. It might give us a *sufficient* cause, but the *efficient*, one suspects here, is to fatten those cows for slaughter.

In a narrative that has run a gamut of infanticide, sexual frustration, pederasty, alcoholism and physical disability, however – a narrative in which Minna, Roley's 'old cow' of a mother, is one of Noah's greatest resentments, a narrative which will end with the castration/murder of uncle Owen, a narrative which sees these very cows, in the passage abovementioned, referred to as 'bloody gins and bitches', and Noah conflate (it is of course a conflation-around-a-conflation) the stink of her own stale-alcohol-and-vomit breath with the taste *in her own mouth* of the dead calves she has had to cut up and pull out of the unready wombs of One Tree heifers – we cannot let explanation stop there, or mask the very real possibility that these elderly cows have in fact been spayed by and in accordance with the needs (/'desire') *of the narrative itself,* a point perhaps not worth making quite so emphatically were that narrative not itself continuous with and synecdochic of the wider discourse – narrative – of its society.

But that – we could call it the *symbolic* instrumentalisation of non-human animals in narrative, their use not only as symbols, but, *as* symbols, as a means of sublimating and so placing out of reach human dilemmas that are seen as irresolvable or simply too difficult to resolve, and itself thus a *symbolic castration/sterilisation* of those animals themselves – while not in any way 'another matter', is perhaps more than this essay can bear.

Another term for foal's bread – *hippomanes*: that which, dried, Roley gives to Noah early in their courtship, to bring luck, and which we find later sitting dusty on a window sill, or a mantel, or hanging from a nail in a shed – is 'foal's *tongue*'. As in the vast majority of 'fine', 'major' novels past and present, the 'voice' of non-human animals in this novel is muted, and the constant suffering of which that voice might tell us therefore substantially occluded. And yet the work, when we adjust our hearing, listen past the 'naturalistic' surface, learn to hear *against* the voice of its narrative, its society, is full of their whispers. So full, indeed, that one might almost think Mears has wanted us to hear them all along.

# Dougald's Goat

## Alex Miller and the 'system ... which as yet has no name'[1]

I'D LIKE TO OPEN WITH THE PROPOSITION that, in a great many narratives, there's a place, a site, where they confess, or at least *pay some acknowledgement to*, the stories they have not followed in order to follow the story that they have. Their *rejectamenta*, their *abject*. And it is not just stories, it is concepts as well, even or perhaps especially ethical positions: sites where they acknowledge some of what has had to be set aside in order for those stories, concepts and ethical positions to come to be. I do not say that they in any way specify or itemise these unwritten stories or concepts or positions, or that their acknowledgement of them is anything but the vaguest symbolisation – indeed, it's so much a matter of the subconscious that it's hard to see how it could be much more – although in some cases they can take a pronounced and almost indisputable form.

In one of the bold philosophical projects of which I've sometimes dreamt I'd in fact go further and attempt to demonstrate a collateral premise, that much of our human ethics are based upon a separation from and rejection – *abjection* is a better term, since this *is* a matter of our identity and what we do to shore it – of the animal, and that it therefore should not surprise us to find the animal (/animals) haunting

---

1    Jacques Derrida, *Of Grammatology.* See note 8.

141

our ethical reflections. Alex Miller's texts, I suggest, are ethical reflections – sometimes profoundly so – and are haunted in this way.

I do not have a name for it – this site or confessional locus that is like standing at the edge of a pit – but for the time being, thinking of Plato's *Timaeus* and discussions thereof by Julia Kristeva (1983) and Jacques Derrida (1993a), I'm inclined to call it *choratic*. I could digress into an account of Plato's concept of *chora*, and even digress, within that digression, into its curious relations to *Cora* or *Kore*, forebear of Persephone and Eurydice, in order to establish an Orphic dimension, but that is subject for a different paper.

If I were to attempt to identify the particular character and strength – the *virtu*[2] – of Alex Miller's writing I would talk first and foremost about its intuitive quality, an *opening* it comes so repeatedly to, not a border crossing necessarily, but certainly a border viewing. As if he were to take us over and again to a figurative door, somewhere within his subject, which, whether or not he or his protagonists accept to do so, the fiction itself then challenges its readers to pass through. This figurative door takes various forms – the transcultural in *The Ancestor Game* (where in fact there is a very literal red door), the ekphrastic in *The Sitters*, etc. – but the intuitive engine, the *propensity* within them is the same. For now, and because it is of particular concern to me, I would like to look at the form – the particularly *choratic* form – it takes in *Landscape of Farewell* and, briefly, in *Journey to the Stone Country*.

*Landscape of Farewell* (2007) is a book about mourning. Each of its central male characters, Max Otto, a German historian, from a university in Hamburg, whom we find on the point of retirement, and Dougald Gnapun, an Aboriginal elder from central/east Queensland, is mourning the death of his wife – Dougald's five years before; Otto's only very recently. And, beyond this deep personal grief, each is involved in a kind of cultural mourning, Otto for a childhood and cultural

---

2   'The soul of each man is compounded of all the elements of the cosmos of souls, but in each soul there is some one element which predominates, which is in some peculiar or intense way the quality or *virtu* of the individual; in no two souls is this the same. It is by reason of this *virtu* that a given work of art persists': Ezra Pound, 'I Gather the Limbs of Osiris' (1911), in *Selected Prose* (1973).

innocence shattered by the activities of the Third Reich, and for the love for his father, a former SS officer, that has been so complicated by it, and Dougald for the suffering brought to his people by white invasion, but also (like Max) for complications thereto in his own family history.

I could be more accurate and specific about these mournings but, although germane to it, they are not my principal focus here. I want, instead, to discuss a particular scene in *Landscape of Farewell*, for which I present the following brief synopsis.

Max Otto has decided to commit suicide. His career seems to him a failure. His beloved wife is dead, and the recent history of his nation and his people has forced upon him a kind of silence and guilt-by-association that have blighted his life. Before he takes his pills, however, he will give a valedictory paper, to mark his retirement, although he feels that in fact it will only confirm his incapacity, since its subject, the continuity of massacre from classical times to the present – the subject of a book he has long wanted to write – has defeated him. He gives the paper – a disappointment, as he had anticipated – and, as he is leaving, to scant applause, a young black woman, Professor Vita McLelland, halts him with a diatribe the only detail from which Miller chooses to give us is that Professor Otto has omitted to mention the massacre of her people.

The applause for her is far louder. Drinks and canapes are served. Max leaves as soon as possible. Before doing so, however, he has a moment of realisation, not just that hers is the voice and perspective of the future, but of something of the extent to which his own generation has let her generation down. He goes to her and apologises. *Landscape of Farewell* could be characterised (all characterisations are partial) as a work of 'Sorry' literature, a literature comprising works by Australian writers attempting the kind of Indigenous redress that for so long their government refused to undertake, and indeed so far has undertaken with little more than word alone.[3]

If this novel *is* such – a work of ' Sorry' literature – then this, Max's to Vita, is the first of its Sorries. Vita is taken aback. Max leaves. She follows, demands that he take her for a drink and, when he declines,

---

3    See, for example, Gail Jones' *Sorry* (2007) and Kate Grenville's *The Secret River* (2005).

assumes, aloud, that he must needs go home to his wife. When he tells her that his wife is dead, Vita is embarrassed and apologises in her turn.

Although it's true that Max Otto says his Sorry first and alone, it's also true enough that Sorries are easier to say when both sides are somehow complicit. It's a minor point, but it does leave something about the nature of saying Sorry open to some further questioning, and so it's important to register it.

Max and Vita do, then, go for a drink, and drink on. She stays, quite innocently, in his apartment overnight. They become friends. She instructs him to come to Australia, to a conference she is organising. She needs his prestige, but mainly, as she tells him, he needs to meet her uncle Dougald.

Max doesn't commit suicide. Although he attends the Australian conference, we are told nothing of that, and next meet him arriving at Dougald's house at Mount Nebo, a mining town in central Queensland.

But now for the scene. How do I describe it? A harrowing passage. A naked woman hanging from the exposed branch of a tree part-way down a cliff behind Dougald's house. A woman, with her tongue blackened, her neck broken by the fall.

Those of you who have not read the book are perhaps suddenly alert. Those who *have* read it are (hypothetically ...) either momentarily confused, alarmed, amused, or perhaps simply annoyed. There *is* no naked woman in that scene! It is Dougald's *goat* who is hanging in this scene. But of course in wanting to *specify*, one in fact *specifies*. Yes, it *is* a goat, a nanny-goat, whom Max, interrupted by Dougald's return after a few days away, has failed to tether properly.

But if you are also thinking '*only* a goat' then I have either failed already in my mission or, more likely (that image of the naked woman might linger), my slow arrow has not yet hit its target. On the wall behind me as I write – the wall of an artists' house lent me by Broken Hill City Council – someone has written out that famous line so often attributed to Theodor Adorno:[4] *Auschwitz begins wherever someone*

---

4    Supposedly PETA (People for the Ethical Treatment of Animals) falsely
      attributed the quotation to Adorno during their 'Holocaust on a Plate'
      campaign (2003). If these precise words can't be found in Adorno, however,
      he makes some statements very like them. See, for example, 'The possibility of

*looks at a* slaughterhouse *and thinks 'They're only animals'*. A bit too heavy for an essay of this kind? I don't know. In my defence I would say that it is, after all, Alex Miller who has written this scene. And it is my contention, of course, that this scene is *choratic*, in the sense just described.

But I am getting ahead of myself. Several pages before this scene there is another, of Max's arrival. A description of Dougald's backyard. In effect it is in two sections, one on page seventy-nine and another two pages later:

He led me back through the kitchen and out onto the square of concrete behind the house. He indicated an enclosed water-tank stand. 'The shower's in there. She's not too bad this time of year.' He turned and pointed towards the back of the yard. A path through the grass led to a wire enclosure in which a dozen or so brown hens and a rooster were penned. Beside the hen run there was a narrow shed constructed of timber slabs with a door at the front. The door of this modest building, like the door of the wardrobe, hung open. 'That's the toilet,' he said. Behind the toilet, beyond the back fence, was an open field in which three large yellow bulldozers, rusting and overgrown with creepers, had evidently been abandoned. 'See them tall trees? The river's down there,' he said, pointing. 'She's not much just now. We haven't had any decent rains this year.' (79)

A ground mist hovered like a softly levitating bed sheet above the open field beyond the hen-run, the abandoned bulldozers a looming family of dreaming pachyderms. All was silent, except for the distant throbbing of the mine. Dougald and I were at the

---

pogroms is decided in the moment when the gaze of a fatally wounded animal falls on a human being The defiance with which he repels this gaze – "After all, it's only an animal" – reappears irresistibly in cruelties done to human beings', *Minima Moralia* (1951). Close enough, one would think, to see the 'quote' as a fair, if opinionated, paraphrase. (See also Susan Witt-Stahl at https://bit.ly/2ZzFige.)

back fence. He had fed the hens and I had collected seven warm brown eggs from their boxes.

'We'd better shift her peg,' he said. His voice caressed the words, as if he spoke in order to listen to himself, in order to hear a human voice in this place. Lifting his hand, he pointed to the freckle-faced nanny-goat. She had cropped almost to the earth the growth of weeds and grasses within the compass of her tether. (81)

These passages and the components from which they are assembled seem innocent enough, perhaps, but in literature there is not a great deal that is innocent (define 'innocent' how you will). Everything is *choice*. Let's start on the periphery – two aspects only – and tighten the focus. The toilet door open like that of the wardrobe in Max's monk's-cell-like room in the house behind them, a fairly familiar image of the proximity of and access to the subconscious, reinforced – I'd say put beyond doubt – by the mention of the proximity and throbbing of the mine. And the way this yard-space has been feminised, even maternalised: the nanny-goat, the hens, eggs warm in the hand, the particular and very significant caress of Dougald's voice (caressing not the goat but the words themselves, as if to emphasise the barriers they represent and create); even the *family* of pachyderms. Put them – the unconscious and the feminine/maternal – together and one might venture the Orphic dimension I mentioned earlier. But that, I think, would be a lyrical seduction, a sleight of mind. There is too much more. Not unsurprisingly for a book which harbours so robust a dream of reconciliation, the passage/space gives on to, for example, the river, site of blessedness, baptism – access, if one can cross it, to the Promised Land (and Miller has mentioned the Promised Land only three pages before). But again I'm ahead of myself. If that is what it does, there is a substantial obstacle to be overcome first.

There are other things here, less appealing and less likely of notice. The hens are caged. The goat is tethered. There are varieties, *orders*, of animals in literature just as there are out of it – the domesticated as pet, the domesticated as food source, the exotic as entertainment, the exotic as source of awe, the exotic as feral/pest, wildlife to be protected, wildlife to be controlled, etc. – and each has its own semiotic, its own *logic* of use and representation.[5] And these, the hens and the goat, are animals of use, as food source;[6] tethered or confined – why should one hesitate to say imprisoned? – accordingly. The scene might be idyllic for Max, for Dougald, and for most readers, but it is doubtful that it is idyllic for those they are at this point looking upon. And, again, if one is tempted to say *but they are only animals* one might reflect that there was a time not so long ago when there were many who, looking on different but related scenes, would have said *but they are only slaves*.

When we come, not many pages later, to the scene first mentioned, of the goat whom Max did not 'properly' tether – who, as she has slipped down the cliff-edge, has been strangled, in fact *hung*, when her rope caught on an exposed tree root, we might legitimately ask just who it is in this book who is most or first attempting to get to the Promised Land? And there are some logical, if conceptually uncomfortable, implications – Dougald as gentle farmer, mournful over but not vindictive concerning the effectual decimation of his race, is also, from this perspective, and I put it as gently as possible, gaoler. The point, and there *is* a point here, is to do with the complicated root-systems of apology – for some, such a clear and obvious thing to do, to say, but for others – how can the white invaders *not* apologise? – well, perhaps that is the door Miller wants to lead us to.

These – the *complications* of this novel as a work of 'Sorry' literature, and perhaps the complications of Sorry literature itself – could well be the subject of a paper by themselves. There is not the space to go far into them here, and yet, as already mentioned, they are deeply pertinent.

Max gives a paper in Hamburg. The paper – its *failure* and *inadequacy* – is the subject of Vita's immediate and devastating attack. The paper, on 'The Persistence of the Phenomenon of Massacre in Human Society from the Earliest Times to the Present' – is on a subject

---

5    There are other animals in this book, for example, even in the scenes we have been discussing: Dougald's beloved and deeply faithful 'pale-eyed wolf-like bitch' (83; on 100 she is 'his sulking bitch': there are some congruities here with the goat), the two younger dogs (83: 'her offspring and members of her tribe') who attach themselves to Max, the bulldozers who, as if pointing up one of the book's redemptive dreams, are caught mid-metamorphosis.

6    Potential, in the goat's case. There is no evidence that she is milked. She would seem instead to serve, as the idiom goes, as a 'lawnmower'.

of profound concern to the book – Miller's book – which follows. And yet, aside from the fact that its list of massacres did not include those of Vita's people, we are given no idea whatsoever of its argument or of the nature of its failure. So too, Max – reformed and rehabilitated, or at least re-*vita*lised, having acknowledged so dramatically his own shortcoming – then goes to Sydney and gives a *second* paper. This, surely, is of even greater relevance, given what has come and is to follow, but this time we are not even given a title, let alone told how it was received; indeed, aside from giving him the excuse to *be* here, it seems to slip entirely from the book's attention. And as to the matter of apology itself, *Max*'s apologies, I will readily admit, are unconditional, and that, of course, is the way they should be. Conditional apologies are not apologies at all. This may be why Miller gives us Max's first apology so quickly and dramatically. But a book, while a plot may unfold within it, is also a *synchronic* thing. Max's apology, as we have seen, is followed almost immediately by an apology from Vita herself, just as, later, Max's second, for the death of the goat, is followed by Dougald's account of the massacre conducted by his great grandfather. *Max* may be able to offer an unconditional apology; *Miller* may be able to offer an unconditional apology; but it seems the *book* needs to balance them somehow, as if one can't happen without the other – although, once again, *why* this must be is largely unstated. And beyond this, as I have already intimated, since it is very much an ethical investigation, the book has an entire ghost- or shadow-economy of the animal, of which Miller seems quite conscious and of which he wants to encourage our own awareness. Again I can't do much more here than sketch: it might be enough to say that Vita's protest, that *her* people don't get into *his* history of massacre, could all too readily be also the goat's stifled cry. There are elisions here, segues between orders of signification, a game – but it is never a game – of snakes and ladders.

There is a feeling, that is, that a number of crucial matters have had to be set aside – to be *relegated* – in order for apology to become possible. This is no surprise, of course: it's been part of our recent national experience. Perhaps it's why the Apology took so long. Perhaps we could say that in *any* situation where what Freud[7] called the narcissisms of minor difference – the *agonising* narcissisms of minor difference – come into play there has to be a great deal of such repression and relegation. Indeed I might go so far as to suggest that in any such process of the reconciliation of two deeply entrenched parties a third party is likely to be involved, but that is, again, to jump ahead of myself.

It is now time to look at the scene itself. To prepare for it one needs merely to remember that Dougald has been away, and that at the moment of his return – interrupted *by* that return – Max, shifting the tether-peg of the goat, has only half-hammered it into its new position. The scene itself occurs in the early hours of the next morning:

> The moment we reached the riverbank we saw her. The ground fell away abruptly at our feet for ten or twelve metres in a near-vertical cliff. It was a dangerous and precipitous place. The elaborate root structures of the great trees had been deeply undermined by erosion, and the mesh of their intricate lattice exposed to the air. Except for a stagnant scum of green algae, which glowed in the cold morning light with a faint and eerie sheen, the riverbed was dry. The exposed tree roots formed the matrix of an elaborate trap. She was hanging by her tether rope, her wooden peg lodged in the fork of a root two or three metres below us. Her tongue lolled from the side of her mouth, purple and swollen, and might have been her disgorged stomach. She hung there, spinning slowly, grinning up at us, her teeth glinting in the rictus of death, her intelligent antique eyes no longer shining with her secret interior life, but bulging blindly, the pupils dull. She was a hideous sight. Her death must have been slow and terrible, for her hoofs had scored the bank deeply in her helpless struggle to free herself.

It is a powerful passage, masterfully written. A brief scan of its imagery tells us what kind of place we have come to. There is the proximity of the river – we have spoken of its significance already – and there is the emphasis placed upon exposed roots that might remind us of Derrida's remarkable, *poetic*, representation of deracination in *Of*

---

7    In *Civilisation and Its Discontents*, for example, where he argues that the human impulse towards aggression is so deep that, in any coming together in *love* there must be a commensurate redirection – scape-goating – of violence.

*Grammatology*.[8] And there is the fact that the dead goat, as placed here, comes *between* Max and Dougald and the river, as if something about her and what she represents threatens to prevent them from ever reaching it (at least until Max *falls with her*, as he does a few pages later, but that scene, and the complications and complicities of it, is, again, an essay to itself[9]).

But although this scan – to reiterate a moment only – may tell us what kind of place we have come to, it does not tell us why. So too a scan – if we can call it that – of the goat herself will give us some intriguing vectors, but will not offer up her deeper significance so readily.

Let's look at her. She is a *nanny*-goat, in which capacity she carries not only the maternal but an additional inflection of the disciplinarian and the teacher (in a sense both Max and Dougald are *infantilised* before her). But she is also mute (her tongue, her strangulation). She is also subject (collar; rope; peg).

---

8   'We know that the metaphor that would describe the genealogy of a text correctly is still *forbidden*. In its syntax and its lexicon, in its spacing, by its punctuation, its lacunae, its margins, the historical appurtenance of a text is never a straight line. It is neither causality by contagion, nor the simple accumulation of layers. Nor even the pure juxtaposition of borrowed pieces. And if a text always gives itself a certain representation of its own roots, those roots live only by that representation, by never touching the soil, so to speak. Which undoubtedly destroys their *radical essence*, but not the necessity of their *racinating function*. To say that one always interweaves roots endlessly, bending them to send down roots among the roots, to pass through the same points again, to redouble old adherences, to circulate among their differences, to coil around themselves or to be enveloped one in the other, to say that a text is never anything but a *system of roots*, is undoubtedly to contradict at once the concept of system and the pattern of the root. But in order not to be pure appearance, this contradiction takes on the meaning of a contradiction, and receives its "illogicality", only through being thought within a finite configuration – the history of metaphysics – and caught within a root system which does not end there and which as yet has no name.' (101–2)

9   As – to introduce yet another of the intersecting loops of interpretation here – may be the fact that she is not actually Dougald's goat – my title is misleading here – but (and although arguably she is no one's) is presented as Vita's, or rather as a goat whose mother has been killed on a country road and whom Vita has rescued, a tale which brings even greater shame to Max and Dougald's subsequently taking so long to bury her.

Clearly she has been killed (or, rather, her death has been *brought about*) – if I read this secret but nonetheless timeless imagery aright – by some mesh in our thought, or perhaps it would be better to say some confusion, some entanglement (we are actually given the word *fork*, and that will do) in the roots that subtend it.

I am not over-reading. Careful emphasis is placed here. We are not just given 'exposed roots': we are given 'The elaborate root structures of the great trees'; we are given 'the mesh of their intricate lattice' ('lattice' is used several times and in several contexts in this book, as if a kind of clue); we are given 'the matrix of an elaborate *trap*'.

She is abject. She is an embodiment of (our) abjection. Indeed, abjection could be seen to be quite consciously adduced – folded back upon itself – in the reference to her tongue appearing to be her disgorged stomach. And she is, of course, and quite literally, a *scape*-goat, of the kind described in the Old Testament, where 'the Lord' details to Moses the process by which animals are to be offered up in sacrifice for the expiation of sin. A bullock for a sin offering, a ram for a burnt offering, and two goats, over whom lots should be cast, to determine which one should be slaughtered, and which one loaded up with the sins of the people and sent into the wilderness, 'unto a land not inhabited' (Leviticus 16.22).

It is, of course, and as already admitted, of the nature of the abject – of this relegation into the Other – to be difficult to name. Is it, in this case, some thing or things we need to set aside in order to think that we, *any* of us, have stewardship, let alone *ownership*, of this or indeed *any* country? Is it that deeply repressed awareness in us that the other side of an apology *for* massacre is complicity *in* massacre, in the forked sense that this book offers us: that *each side*, Indigenous and invader, has engaged in human massacre, and that in forgiving – apologising to – each other, they are glossing over the fact – awareness – that even as they do so they are still engaged in, are mutually complicitous in, that vastly wider massacre that Isaac Bashevis Singer has called Eternal Treblinka?[10] Is it that great paradox of our deeply divided behaviour, which I have elsewhere called our wound:[11] that, in order to be

---

10   In his story 'The Letter Writer', *The Séance and Other Stories* (1968).
11   'The Smoking Vegetarian', above.

merciful, we have to *turn our backs* upon our continuing mercilessness; that, in order to be compassionate, we have to *turn our backs* upon our continuing lack of compassion?

The nanny-goat's throat is constricted; her tongue swollen and purple: she is an image at once of the voiceless and, perhaps, of the unspeakable. My point is less to attempt such specifications of the abject than to point out a site – a site, and perhaps a process. *Choratic*, I have called them, though others may wish to use other terms. And, of course, to encourage a kind of reading *for* such sites and processes. Even when they cannot be found – even if my opening premise is a kind of wild hope and exaggeration – it is surely incumbent upon us to look for them since doing so galvanises the issue of the unspoken, which is not, after all, so much the unnecessary of our lives (and narratives, and ethical positions) as the very matter from which those lives and positions have been carved.

In case my reader might be inclined to think such findings contrived or fortuitous, I'd like to glance, in closing, at another novel of Miller's, *Journey to the Stone Country* (2002), and a similar choratic scene, in Bo Rennie's account of the scrub bulls of Zigzag station:

> 'These ridges are full of old scrub bulls. They eat them poison zamia nuts when the feed cuts out in the winter, then when the heat comes on in summer the rickets come out of their limbs. They go down in the hindquarters and get themselves snared up among that shattered basalt. When we'd ride up on one of them, the wild dogs would be sitting in the shade close by, watching him die, taking it in turns to jump in for a quick bite every now and then ... I've lain there plenty of times in my blankets in the moonlight listening to one of them trapped bulls bellowing. That high-pitched bugling sound they make. You'll know it when you hear it. The trumpet of the angel, my dad used to call it. Carries all up and down these ridges ... I used to lie there at night thinking that old angel was out there turning the stones over looking for me.' (138)

Eloquent enough by itself, the passage becomes even more so, and more destabilising to the ostensible ethics of the novel, when we realise that the Townsville house of the parents of Annabelle, Bo's white co-protagonist – the house in which Annabelle is temporarily staying – is on Zamia Street, that a whole chapter takes its name from this street, and that a chapter soon to follow, curiously titled 'A Plague of Dogs', contains the following lines – and oblique acknowledgement:

> 'My dad used to tell us you people fought like wild dogs over there at Verbena.'
> 'Wild dogs? Well we fought when we needed to fight, that's what we did. We had some real good fights out in front of that big tamarind tree of Grandma's.' (226)

But, to the sceptical, even corroboration needs corroboration. Look then at the curious paradox, elsewhere in the book, by which, although they note the roadkill, even count it, as if in sympathy for the creatures thus slaughtered, Miller's own protagonists – the heroes of his tale – drive the country roads carelessly and at high speed, even at dawn and dusk, the peak danger times for wildlife, doing little or nothing to avoid contributing to the toll.

This paradox seems to concentrate, if not in any way resolve, much of what at a conscious or unconscious level Miller seems to be wrestling with in these texts, the way the ethical resolutions we struggle for, while other, far deeper and far more difficult business remains unaddressed, will seem like guilty rhetorical platitudes, matters less of forgiveness than mutual complicity, fragile, vulnerable not, like our other human constructions, before any tsunami or flood or earthquake – those nice available metaphors – but before something far more damaging to our own senses of ourselves, the return of which, like the return of anything too long repressed, can be devastating. I'd like to call it our own deep moral conscience but I'm not sure that wouldn't in itself be a platitude. In truth I don't know its name, or whether it even has one.

Let me try to approach it from a different angle. In the speeding/ roadkill trope just adduced you could be forgiven for thinking that Miller looks at his own characters with a measure of scepticism, even disappointment. This disappointment and scepticism – their *complications* – return in the novel's closing pages. Bo takes Annabelle to see Panya, an old Aboriginal woman with first-hand experience of a massacre in which Annabelle's own grandfather participated (the grandfather who, in his senility, befriended – became the constant companion of – a bull).

Disgusted that Bo has brought Annabelle into her house, Panya unleashes a diatribe of truth and of horror that neither Bo nor Annabelle have any answer for, and leaves the reader uncertain as to how, at last, to regard them. Certainly any *ethical* resolution they represent – and I do think they are intended to represent an attempt at one – looks hollow indeed, based upon a turning one's back upon, rather than a facing of the past. Yet these are also characters lauded – beloved – by critics and author alike. What is happening here? It is as if a door has been left open – Panya's door – and a chill, choratic wind is withering the human landscape. Where has it come from? Panya? But she herself is victim, not origin. Even her message is a messenger.

To get the beginnings of a better idea, perhaps we must look at how it was that Panya, as a young girl, survived the massacre in the first place:

> Your grandmother's old lady hid us two kids with her in the hollow carcass of a old scrubber bull that was laying out in the open … Me and your Grandma was all curled up inside that carcass looking out through the old bull's skullholes watching them men murderin our people in the moonlight. (338–39)

Then go one rather obvious step further – to see this world through a non-human animal's eyes, and to realise, perhaps, the irony here, that while one grievously neglected massacre may have been added to Max Otto's list, another, the 'Eternal Treblinka' of non-humans, is still waiting.

# An Exoneration

## The Grieving Kangaroo Photograph Revisited

Photo by Evan Switzer.

ON 13 JANUARY 2016, the *Fraser Coast Chronicle*, a small Queensland newspaper, published, with a sequence of four striking photographs, a piece by Amy Formosa entitled 'Photographer captures kangaroo family's grief'. The photos, taken two days before, 'went viral', and were viewed around the world.

One image in particular stood out: an adult male kangaroo cradling the head of a dying doe – lifting her, or so it appeared, so she could look more directly at the joey standing before her. Her arms, in this photograph, reach out as if to catch the joey's forepaws one last time. It was hard not to think of this as a family. It seemed, contrary to what we're led to believe of mob structure (a dominant male, 'servicing' all does), as if this buck was the doe's partner and (this said more confidently) the joey their child. Grief and bewilderment – the pain of loss – well from the picture. I can only think it was shared so widely because of what people *recognised* in it, an emotion that crossed the species barrier.

The photographer was Evan Switzer, and the photographs were taken in open parkland at River Heads, Queensland, a coastal settlement three hundred kilometres north of Brisbane. Switzer had been exercising his dog, a brown Labrador female. He would later tell me that it had been she who had found them and brought them to his attention. The kangaroos there were used to her, he said: they'd seen her several times before. I mention this because I think it's significant it was a non-human animal who brought our attention to the suffering of another non-human animal. Albeit silent, and resting most of the time, her presence shouldn't be overlooked for the sake of an essay's convenience. When he realised what was going on Switzer walked back to their house with her to get his cameras and her lead, then drove, with her, back to the kangaroos. They parked some distance away. He attached one end of her lead to the car, walked as far towards the kangaroos as her lead permitted, then photographed from there. The kangaroos, he says, were about ten metres away.

It's impossible from this photograph to tell *why* the doe is dying, or even *that* she is dying: she may be dead already. Switzer says he could see no signs of injury, but River Heads is a popular departure-point for Fraser Island and there are cars about. One strong possibility – the photographs convey the sense that this death has been sudden – is that she's been hit by a car and has sustained fatal internal injuries. Very often such injuries are not readily apparent to an onlooker.[1]

We might leave it there – a powerful image, or *set* of images (a total of six photographs were eventually released), offering a glimpse into the human-like emotional life of a non-human being; something that, as such epiphanies can do, brings us somehow *closer* to those beings, and them to us, suggests a kind of common ground. But I'm not sure we *can* leave it there. Even the image of a dying kangaroo, it seems, is political.

It's unclear whether the *Chronicle* took these photographs to an expert, or was simply reporting on that person's independent response to them, but the next day, hot on the heels of the publication of this piece, came another, by Lea Emery, entitled 'The ugly truth about that "grieving roo" photograph', conveying the opinion of a man I'll call 'Dr E.', emphatically dismissing the initial interpretation of the image.[2]

By this time the *Chronicle* piece had been picked up internationally. Both *The Daily Mail* and *The Guardian* carried versions on 14 January, as did *The Washington Post* and many other newspapers. Evidently the emotional impact of these photographs was powerful and extensive.

It appears that one or another of the syndicating newspapers – I think *The Guardian* but these things can be hard to track – had then approached an expert of their own, from a reputable Australian university. Almost immediately, in any case, *The Guardian* published its own retraction, presenting the report of this expert.[3] 'Dr [S.]', it began, 'a senior lecturer in veterinary pathology, says it is "gross misunderstanding" to think [the] kangaroo was cradling [a] dying mate'. 'The photographs', it explains, 'showed the male kangaroo "mate guarding" – holding other males at bay':

> 'Competition between males to mate with females can be fierce and can end in serious fighting,' [Dr S.] said. 'It can also cause severe harassment and even physical abuse of the target female,

---

1    See the photographs and original *Chronicle* piece here: https://bit.ly/3iu1eR0
2    https://bit.ly/3mmRryS
3    Elle Hunt, 'Kangaroo in "grieving" photos may have killed while trying to mate, scientist says' (14 January 2016): https://bit.ly/32rDZl5

particularly when she is unresponsive or tries to get away from [an] amorous male.

'Pursuit of these females by males can be persistent and very aggressive to the point where they can kill the female. That is not their intention but that unfortunately can be the result, so interpreting the male's actions as being based on care for the welfare of the female or the joey is a gross misunderstanding, so much so that the male might have actually caused the death of the female.'

It's a curious statement, somewhat illogical in its construction. How is it that a set of *possibilities* – 'Competition … *can* be fierce', '*can* end in … fighting', 'Pursuit … *can* be persistent', '*can* kill the female', '*can* be the result' (emphases all mine) – can become, without any connecting explanation or evidence, a *fact* ('*so* interpreting … *is* a gross…')? How can it be that this misunderstanding can be *such* a gross one ('so much so') that *it*, that is, the misunderstanding (for this *is* what the sentence says), 'might actually have caused the death of the female'? *It*, and not an illness, not being hit by a car, and not – paradox of paradoxes – 'competition between males' or sexual aggression? One shouldn't make too much of this – it's a simple grammatical slip, and very likely the newspaper's, not the expert's – but the fact remains: the statement, as printed, effectually *undoes* itself.

Dr S., the article continues, 'added that, though he thought the term was often misunderstood and misused, the reporting of the viral photographs had been "naïve anthropomorphism".'

'Naïve anthropomorphism'. Count to ten, when talk of an animal's emotions is concerned, and, like the proverbial bad penny, *anthropomorphism* turns up.

Defined by the *Oxford English Dictionary* as 'the attribution of human personality or characteristics to something non-human, as an animal, object, etc.', *anthropomorphism* has long been employed to criticise those who argue for the presence of human-like emotions and emotional intelligence in non-human beings, with the implication that such attribution is a kind of intellectual embarrassment, an *error of thought*, a naïveté to the point where a term like 'naïve anthropomorphism' becomes virtually tautological. From an animalist perspective, on the other hand, the accusation of anthropomorphism can only appear as a means of hosing down any sparks of empathy and identification with non-human animals before the species barrier begins to smoulder.

This isn't to say there's no such thing as 'naïve anthropomorphism'. As with a great many of our concepts, from 'the survival of the fittest', say, to 'deconstruction', there can be useful, sophisticated and comprehending forms, and there can be simplistic and naïve ones. People will dress their cats and dogs in human costumes, insist their tortoises have opinions about television programs, or claim their octopus can predict the winner of a football match – that is, will force a demotic humanness onto non-human animals in a manner that demeans non-human and human animals alike – but blaming anthropomorphism per se for such naïveté is itself an intellectual error.[4]

'The kangaroo's "sinister" intentions', *The Guardian* continued, reverting to the expert quoted in the second *Chronicle* piece,

> were first flagged in an explosive blog post by Dr [E.] … of the Australian Museum. He praised Switzer's 'great photos of the kangaroos', but said they had been 'fundamentally misinterpreted'.
>
> 'This is a male trying to get a female to stand up so he can mate with her,' he said.
>
> He pointed to the 'highly stressed and agitated' state of the male kangaroo, which had been licking its forearms to cool down. [Dr E.] also pointed to 'evidence … sticking out from behind the scrotum' of the kangaroo's sexual arousal.

From a grieving partner, the buck has become, overnight, a 'sinister' killer (*The Washington Post* actually uses the term *necrophilia*). If the original piece presenting the photograph as of a grieving kangaroo went viral, the 'revelation' of its 'misinterpretation' went far more so. Search 'grieving kangaroo' even now and you'll find many more pieces gloating over the gullibility of those believing the buck was grieving than accounts suggesting he was actually doing so ('No, this kangaroo

---

4    For more on anthropomorphism – and some inevitable reiteration – see 'Writing Animals' below.

wasn't grieving – it was raping a dead female' [Harry Readhead, *Metro News*, U.K.], 'Grieving kangaroo actually just wants sex' [*Xinhua*, P.R.C.], '"Grieving" kangaroo photos may actually show brutal murder scene' [Pardes Seleh, *Daily Wire*, U.S.A.], etc.) We've been hoodwinked – tricked by our own sentimentality – if we think this apparently tender, traumatised kangaroo is anything other than ('the "ugly" truth') a sexually over-charged male frustrated at having (René Descartes here: the non-human animal as *machine*) busted his toy before he could have his way with it.

Whether we see this as trial by sensationalist media or by a public responding to fodder that media had served them, the buck had in effect been charged with and, unrepresented and on the slimmest of evidence, found guilty of a heinous crime. If one reflects, with sad irony, that there's a name for this kind of court, and understands the term more deeply and ashamedly than ever before,[5] one reflects just as ruefully upon the alarm and denial with which so many view any opening of the door to the possibility of human-like emotions in non-human animals – a door which, opening, exposes the extent of human cruelty and their own unwitting participation in it – and the vicious pleasure they take in slamming it shut.

Is there any way the buck might be exonerated? Where would one begin? The 'experts' have certainly stacked the cards against him.

Tempting as it might be, in a time of widespread and indisputably justified condemnation of abuse by rogue *human* males, to turn the accusation of anthropomorphism back on the scientists and suggest that even their own supposed objectivity might reflect and unwittingly exploit some very topical human preoccupations, I think we do need to do some further thinking here. Dr E. has drawn attention to a part of the image most of its viewers won't have caught. Is he correct in his interpretation of what he's found? Should we now pack away, embarrassedly, our felt connection with this (as we thought) family group, or is there more to the issue?

---

5　A *kangaroo court* is defined by the *Oxford English Dictionary* as 'an improperly constituted court having no legal standing', and by Wikipedia as 'a court that ignores recognized standards of law or justice, and often carries little or no official standing in the territory within which it resides'.

Kangaroos (for example) are thickly supplied with veins on their forearms, close to the skin's surface, and cool themselves, on hot days, by licking that surface (and that of their legs, belly and tail). While it may be true enough, as Dr E. points out, that the fact that the buck, in several of the photographs in question, has been licking his forearms is an indication that he's 'hot and bothered', he's happy to leave this as an indication of sexual arousal and doesn't mention that on 11 January 2016 – the humid midsummer Monday on which these photographs were taken – the temperature in River Heads was 30 degrees in the shade and a good deal higher out of it, warm enough for a kangaroo to have been licking his/her forearms quite simply to cool down. Nor does he mention how clear it is from the published photographs that the joey, also, has been licking his forearms. Must we think of him, too, as part of this supposed sexual arousal and aggression, towards his own mother?

To do Dr E. justice, it's quite true that in some of these photographs one glimpses the male kangaroo's erect penis – a more reliable indication of arousal than licked forearms – but of what, if anything, is this conclusive? Chimpanzees in a zoo masturbate when excited or distressed; elephant males can develop erections when distressed; castrated rams can develop erections on hot days; lazing dogs can develop them when there's no other dog for miles; stress or anxiety can cause otherwise unwarranted erections in human males; the penis of the kangaroo male 'is extended erect while eating',[6] and, preparing to write this essay, I've found several photographs of such males, in what I call the 'guard' position, with erections.

Still, although it doesn't seem to me the presence of an erection, in such a stressed and distressing situation, must only and necessarily be a sign of sexual arousal, we can't dismiss the possibility. Must we even then, however, interpret it in the manner this expert has? What do we know of the inner workings of a kangaroo's psyche? I can think of at least one play by Shakespeare (*Othello*) in which sexual excitement, murder and love are deeply confused in the one person and moment. And over and again, reading yet another of the Australian Broadcasting Commission's all-too-frequent reports of a domestic homicide, one

---

6    Staker (2014), 2.

Photo by Ray Drew.

wonders whether this or some such lethal cocktail might have been involved. Which, though it might point up, ironically, the anthropomorphism in the accusation itself, is not to imply the male kangaroo in the photograph *has* killed the doe.

Let's nevertheless entertain Dr E.'s hypothesis that the buck might have accidentally killed the doe in the process of a sexual encounter. Would even *that* mean he could not be dismayed, horrified, confused (etc.) by what has happened – that he cannot *grieve*? Why would even this most extreme case mean that, as the experts say so emphatically, this is *not* a photograph of a grieving kangaroo?

There are, moreover – to cite only the most obvious of reservations here – *three* kangaroos in the photograph. The doe, if not dead already, is dying, and in this extremis may well be experiencing her own form of grief, as is (I think we can surmise) the joey. To say that, because *it may be* that the *buck* may not be grieving in the unalloyed, uncomplicated manner in which people first looking at the photograph might have assumed, this photograph is *not* a photograph of a grieving kangaroo, is quite a stretch (and a tad chauvinistic).

Why – a second reservation – must we *normalise* grief, and then impose *that* grief-model upon non-human animals, if only in the negative form of denying their capacity for it? For all the attempts of a white, Western intelligentsia to universalise their own perspective,[7] humans themselves do *not* have the one consistent mode of expressing grief. There are humans who place their deceased in vats until all juices have leached out. There are humans who sleep with their deceased partners until the rotting flesh drops from their bones.[8] Zoroastrian grief, it would seem, accommodates placing the dead on especially constructed platforms to be eaten by vultures. Why must we deny *non*-human animals grief-modes of their own? Isn't it just possible, for example, that this male kangaroo feels that sex with his partner might actually revive her? In denying him *un*known alternatives of motivation, and imposing only those they *know*, aren't the 'experts'

---

7    Are we talking about human grief in general, my partner asked me, or are we talking WEIRD (White, Educated, Industrialised, Rich, Democratic)? See Heinrich et al. (2010).

8    Metcalf and Huntington (1978).

providing us, ironically (and yet again), with an example of the self-same anthropomorphism of which they accuse others?

ROLAND BARTHES (*Camera Lucida*, 1980) speaks of the *punctum* of a photograph, that detail which seems at once to jump out at us and to centre the image somehow – a point which not only commences and guides our 'reading' of that photograph, but makes us want to read in the first place.

The term can be applied more broadly. At a time when the kangaroo, as a set of tribes amongst us, is so widely and cruelly persecuted, the very *state* of its *race* a grieving, this particular photograph could be seen as deeply symbolic, the *punctum* of a much larger picture. Day after day, night after night, in almost all parts of the country, kangaroos are being shot or run down, their families, mob structures and cultures destroyed, to the point, in numerous areas, of regional extinction. Accept *this* grief, *as* such, and we risk – nurture – a vast and devastating realisation, with consequences not only for our grotesque persecution of the kangaroo as a species, but for our relations with non-human animals, period.

What *is* the *punctum* of this photograph? Is it in the eyes of the buck, looking directly at us, his face with its complex expression of alarm, bewilderment and deepening sadness? Strangely, given that the photographer was only a few metres away, it doesn't seem an expression of warning or annoyance (but how could we know? It would be *anthropomorphic* merely to look for such things, let alone find them), although, *were* he in the 'mate guarding' position, fending off other males, it would seem logical to think we'd find a trace of those emotions.

Is it, instead, his paws about the neck of the dying doe as, in one interpretation, he holds her head up to see her son one last time or, in the other, grasps her in attempted possession? Is it her arms, reaching out, as they seem to do, to her joey? Is it – something quite different – the two grazing kangaroos in the distance behind them (oblivious, like the ploughman in Breughel's 'The Fall of Icarus')? Or is it – something at once more subtle and more striking – the image's extraordinary resemblance, in its configuration, to Michelangelo's *Pietá*, that iconic depiction of (Western/Christian) grief? The answer may vary from viewer to viewer. There may be more than one *punctum* in the first place. It may even change as one's reading develops.

For me, the extended arms of the doe come close, or rather the paws. Or did, initially. That it's both of them, just touching those of the joey, and – although one knows this may just be *rigor mortis*, an indication she's been dead an hour or more – that they're so tense, every digit stretched to its maximum, as if in a last, anxious effort. Though I'd not dismiss the two kangaroos grazing in the distance, for the emptiness they seem to emphasise. The 'experts', who insist that this is not a photograph of grief but of a male clutching, in a manner designed to ward off other males, a female he may just have killed in the process of coition – have they taken this emptiness into account? Where *are* the competing males? If, as the photographer, interviewed for the first *Chronicle* piece, does indicate, there *were* other kangaroos close by at some point, and this buck chased them off, they don't seem to have returned. Certainly they're not in any of *these* photographs.

There is also, of course, the matter of the photographer himself. The photographer and his dog. Might their presence contribute in some way to what the photographer sees and now shows us? Have the buck and joey been *dislocated* from the doe somehow? Has their grief been disrupted? That impossible question – what might there have been to photograph, had the photographer/dog not been there? – must still be asked. Mightn't it be that there are *two* events in question here, the death of the doe, and the encounter between buck, joey, photographer and dog, and that we can't really interpret one without the other (Switzer has stated, for example,[9] that, far from acting aggressively towards him, the buck seemed almost to be *asking for help*)?[10]

In several of these photographs the buck looks directly at the camera. And, in several, he is supposedly in the 'mate-guarding' position. Could he be guarding his mate *from* the photographer and/or dog? Could this event have been extended (the photographer says they were there for an hour and a half) *because they were there*?

---

9    Telephone conversation, 5 October 2018.
10   I note, late 2020, publicity recently given to a combined UK-Australian study determining that kangaroos have the ability to ask humans for help. See McElligott, O'Keeffe and Green (2020).

The buck is unable to move the doe. Strong as he might be, he's not built for such things. Even if he could it may be he'd not attempt to do so while the human and his dog were there. And the human, aware that what he's seeing is so extraordinary, will in his turn stay as long as he's able. Each of them, in other words, is caught up in something, an event, an opening – a psychic, perceptual, inter-species phenomenon – which, random as it might seem, is larger than themselves, an event none of the participants at this stage can have any real idea of. Through its one, key image, this event will at one and the same time act as a portal connecting – or offering the *potential* of a connection between – the human, dog and kangaroo. It will reveal, for example, the kangaroos' sentience, their capacity for human-like feelings, their capacity (let's say) to experience the stuff of tragedy *as* tragedy.

Depending upon the manner in which this image is interpreted (whether as 'grief at the death of a mate', or as the product of 'a murderous sexual drive'), it will either sensitise its viewers *to* the plight of kangaroos (and their *like*ness, their *kindred*ness) or confirm a disastrous *de*sensitisation. This kangaroo has become, in this event – through the publication and massive distribution of this photograph – all kangaroos. Our interpretation is therefore crucial. The tiniest sliver of information may prove pivotal.

A PHOTOGRAPH FREEZES a moment in time, and in so doing also separates itself from time, or changes its relation to it, preserving indefinitely what might otherwise be gone in a second. This photograph may be around for a long time. All the more reason, surely, to get its reading right, or at least attempt to rectify a brutal *mis*reading. But, for any photograph, there is also a *before* and an *after*, and the photograph's moment, isolated as it may be, is heavy with them. This photograph in question does not stand alone. In all, as I've explained, five others have been made available to us.[11] Could I place them in order? Might this help me better understand what has gone on here?

Three of them form a tight sequence, although in which direction this sequence runs isn't so clear. It either shows the buck lifting the doe's

---

11   Largely in the original *Chronicle* piece, and in the first *Guardian* piece at
     https://bit.ly/3hzum8w

head and shoulders (to the point of the photograph about which our discussion has centred), or gently placing them on the ground. That these three images do form a sequence is confirmed by the presence, at the same place and in the same postures, of the two grazing kangaroos aforementioned. In each of these three photographs, it's also worth noting, the forearms of the adult male and the joey are slick with saliva.

The joey, in the photograph in which the doe is half-raised, leans in towards her, and in the photograph in which she is at ground level bends down, I think to sniff the opening of her pouch, although whether because, clearly an 'at-foot' joey (too large to return to the pouch), it has, until recently, been his home, or because he's concerned for the fate of a further, younger joey there, is not clear. *Is* there a second joey? Did anyone – the photographer, others walking through the reserve – think to check?[12]

In a fourth photograph – I'm not suggesting an order here – the buck and joey stand apart, the doe lying between them. In a fifth the buck stands high, in the alert/guard position, with the joey beside him and the doe at their feet. In a sixth the joey stands a few paces to the right and the buck, beside the doe, leans over her, nuzzling or sniffing (one might almost think kissing) her face. Photographer aside, you could construe this as a deeply private moment, a loving farewell.

It's a puzzling photograph. The doe's head is now to the left. Since everything around her remains substantially the same, one has to assume it's the doe whose position is different, and that either someone – the photographer? the buck? – has moved her, or, still alive, she's somehow moved herself. I thought, when I noticed this, of the circles left in the mud or gravel or grass beside a road where a kangaroo whose leg has been broken in a collision with a vehicle has tried to get up but, because only one leg is working, can do no more than turn themselves around and around. Is that what's happened here? Has *she* turned herself around?

Some *ifs* would have to line up here if this were to be pertinent. *If* she's been hit by a car, the chances are strong she'd have an injured leg, and *if* she has not (yet) actually died, then it's possible she's at some

---

12 The photographer has subsequently confirmed that he did check her pouch (see below).

point struggled to right herself, and so moved in the direction in which the photograph suggests, but I think it far more likely she's been shifted by the buck's attempts to right her. It would help if we could establish a particular sequence here, but that seems impossible. I might have tried to contact the photographer and ask him about such things, but up to this point it had seemed – if, to exonerate the buck, I wanted to work with the same materials the general public had used to condemn him – a kind of bad faith to do so.

Both sequence (could we establish it) and *punctum*, in any case, in drawing attention to themselves and setting the register of a reading, can actually distract one from other and perhaps more significant details. I'm embarrassed to think how long it took me to contemplate the *legs* of the doe. Was it just me – was I dramatically underestimating the foreshortening caused by the perspective – or was I right, at last, in thinking her right leg, the lower part of which was angled, awkwardly, directly towards the viewer, had been distorted somehow, maybe even severed somewhere above the last joint? Even if it *was* just foreshortening, and she was quite intact, the angle of that last section – her foot – seemed exaggerated, as if her leg had been badly broken or was being held in an unnatural position.

It was only now I felt I could and perhaps should contact the photographer. Although it seemed to me there must be an injury, no amount of enlargement and close examination of the photographs available could confirm it. The buck was still not entirely in the clear. Switzer is on record as having stated he could see no signs of physical injury upon the doe. I needed to find out how closely he'd been able to examine her, and whether there was any chance he'd missed something.

In the first of our telephone conversations (5 October 2018), while repeating that he'd seen no injury, Switzer told me he'd been able to approach the doe only once, to check her pouch at a point where the buck and joey were standing off at a little distance, and that, while he did so, he'd been too concerned to watch the buck (who, it must be said, would almost certainly have attacked, if he *had* been sexually aroused) to check the rest of her, and so could neither confirm nor rule out a leg injury.

As to her changed position, no. He'd watched for over an hour. At no point had he seen her kick out or in any way move independently. The buck, on the other hand, had attempted to lift her so often that her position might well have changed during the process. To aid my thinking, and on the chance there might be a photograph that showed an injury more clearly, he sent me his full set of just over two hundred, all taken in the eighty minutes from 4.36 until 5.56 on the afternoon in question.

Although difficult to quantify exactly, the buck, in these photographs, lifts or attempts to lift the doe's head and shoulders between thirteen and sixteen times. He has an erection in almost all of these attempts, but, significantly, there's no evidence (at least not that I can find), in any of the images, of attempted intercourse. Instead most of these attempts to lift her could be seen as attempts to hold, and perhaps rub, her head against his chest.

'Chesting' is a common behaviour in Eastern Grey kangaroos. The kangaroo chest contains powerful scent glands. These are used to mark territory, although the question of marking raises the issue of the complex and ambiguous relations between endearment, attachment and possession. The buck's attempt to rub his scent onto the doe – if this *is* what it is, rather than simply to hold her close – may be as much to protect her (or her body) from the encroachments of other males as to establish her as his territory – may be a matter of love or affection rather than maintenance of position. Curiously, although we might expect it in a sexually aroused male, very few photographs (there are some) show the buck sniffing at or paying any particular attention to the doe's hind quarters/cloacal area.

Considerable attention, on the other hand – seen in seventeen photos – is paid to her *ears*, and in a dozen more such close attention is paid to her face and mouth one might almost (again) construe it as kissing.

Overall, the weight of this extended sequence seems more on the side of grief and endearment than attempted coitus, the erect penis rather, and as already suggested, an indication of distress than sign of sexual intent. All very well, but this buck has also been accused, if not of actual murder, then of the kangaroo equivalent of manslaughter. Only unambiguous evidence of an alternate cause of death can begin to absolve him. Do we have it?

At first I thought not. The vast majority of these photographs are taken from the same position as five of the six already published, and

Photo by Evan Switzer.

therefore show the same foreshortening of the right leg without in any way clarifying the cause of that distortion. There is one, however, in which the buck, attempting to roll the doe onto her back, exposes her right leg quite clearly. The ankle, in this photograph, appears badly swollen and out of alignment, and the leg at this point unnaturally bent, in contrast to the stiffened and outstretched left leg.

My knowledge of macropod anatomy is limited. I sent both the full photograph in which the leg's shape clearly appears, and an enlargement of the ankle itself, to two very experienced wildlife rescuers, each of whom has an intimate day-to-day knowledge of kangaroos.

Each confirmed, almost instantly, that the doe had a displaced fracture of the right ankle. And each suggested (something of which I'd made no mention) the most likely cause of her death was that she'd been hit by a car. One, commenting on the ankle, wrote:

Photo by Evan Switzer.

*The leg is like a spring. It has to take a lot of force on it. The movement of the legs keeps the heart and lungs functional. Sorry to say, keeping her off the leg, her heart could not take it.*

I then asked one of them whether there were any scent glands in or near a kangaroo's ear, since the buck had been paying the doe's ears such attention. He said he was unaware of any scent glands there, but that roos who were close to one another would often groom one another's ears.

Death, then, from a combination of external and internal injuries, after a collision with a motor vehicle, or from heart failure as a consequence of the same. Or, perhaps, shock from a loss of blood. Ultimately it's hard to say. Switzer has reported (21 May 2020) that there was a man slashing grass on the reserve on the day in question, and that this man later told him he'd seen 'quite a bit' of fresh blood near the reserve's entrance and in the grass nearby. Very likely this blood was hers; it might even indicate the site of the accident; but, so late in time, this must be regarded as circumstantial evidence only. Some sense of the freshness of the wound – whether that dark area on its underside is dried blood, fresh blood, or not blood at all – might have helped, but it's too late for that too. Switzer doesn't recall seeing blood as he checked her pouch, but, as we've seen – having to check so hastily – there's an explanation for that. There *are* some curious dark patches on the buck's hide. These may be blood, and from her, but there are many ways he might have come in contact with it and I can't see that this implicates him.

Can I say, *categorically*, that there's been no 'rape' of a dead female, that this *isn't* a 'brutal murder scene', or that the buck has *not* 'just wanted sex'? No, of course not, though on the one hand if the buck 'just wanted sex' I can't imagine he'd have hung around so long and so fruitlessly, and, on the other (albeit a little paradoxically), that he *has* hung around so long seems to indicate that he has *not* had sex, and therefore to go some way towards absolving him of his remaining charges.[13]

My strong sense, given the weight of circumstantial evidence, is that he's not guilty of any of these things, and that, if there's been any bestiality here, it's been in the viewers, not the viewed – that a massive injustice has been done to him and, through him, to his species, and that we owe them extensive redress.

A strong sense, however, is not a fact. Maybe this essay and I have reached a point where we have to acknowledge our own irresolution. But maybe, too, this irresolution is a point in itself.

As the common root of the words *ethics* and *ethology*[14] suggests, ethics are species specific.[15] To impose the ethics of humans upon other creatures – ethics we ourselves find difficult to maintain – may itself

---

13  It seems clear, from the repetitiveness and uncertainty of his actions, that the buck is confused. One very plausible suggestion in this regard has been that he can sense, by smell, that the doe is ready for sex, but can get no response from her, and that it may be it's only as he attempts, repeatedly and unsuccessfully, to engage with her, that he realises that, contrary to the message her smell is giving him, she is dying or already dead. While I'd reject any implication that a kangaroo cannot recognise death, I could accept that it may be that, the death being so recent – perhaps occurring mid-sequence, as it were – the signs which would indicate this to him have not yet, or (at some point in this sequence) only just become apparent. This may have implications for the interpretation of any particular photograph in this sequence as of a *grieving* buck kangaroo, but only in as much as it may now be an issue of when puzzlement, dismay and confusion give way to realisation, and grieving starts.

14  The study of (non-human) animal behaviour.

15  As I've argued elsewhere. See *The Grass Library* (161–2).

be unethical. We live in houses and apartments. We drive cars. Our towns and cities are built to our amenity. We've little idea, from within such amenity, of the harsher realities of the non-human creatures we live amongst, or the particular ethics those realities impose. Yet, when it suits us, we impose our ethics upon them, and blame, dismiss or punish them accordingly. The recognition of this as a manifest impropriety, however, does not mean our own ethics, with regard to ourselves, can be suspended. For all the attendant difficulties and incompatibilities of our relations with our non-human neighbours, we have, towards them, as far as our *own* behaviour is concerned, an ethical responsibility.

I have spoken of a bigger picture, and of this one photograph of a grieving kangaroo as emblematic of the grief and suffering of kangaroos as a species. A part of this larger picture, as I've also stated, are those who would wilfully impose that suffering in the first place. It would be foolish to imagine that the rapidity and brutality with which the welling of understanding, empathy and compassion this photograph engendered was repressed was not in some measure political; more specifically, that it wasn't in part driven by proponents of the kangaroo 'industry' which was, at that particular point in time (and continues to be), avidly seeking to expand the international market for its products, kangaroo meat in particular, and assure those markets of the sustainability of their supply.

Empathy for kangaroos does not help that project; assurances of their violence and bestiality, on the other hand, have obvious benefits. It is easier to persecute those whom one has first classified as bestial (or as pest, as is also alleged). If, apart from a fair amount of logic and common human sense, we have no more concrete proof of this buck's innocence than those who have accused him have of his guilt, it would seem to me that it is *we* who are left with, and must make, a moral choice.

The final images in the sequence show the buck alone with the doe, now clutching her to his chest, now laying her down, now lifting her again, unable to leave, unable to stop holding. The shadows are lengthening. The roos who were grazing in the distance have moved away. The joey has wandered out of frame. One lone kangaroo, on the far side of the wide, open space, looks on. Ungrounded, or so it would seem to me, utterly lost, the buck, in the very last photograph, looks off and downward at something that may not be there at all. The eyes of the

Photo by Evan Switzer.

doe, if we look closely enough – through the camera lens, through the years, through whatever barriers we might try to put up against them – are staring directly at us.

# Writing Animals

THERE ARE MANY FORMS OF 'ANIMAL' WRITING – one might write of farming animals, or of hunting them, one might taxonomise, or write a history of our relations with them, or of our representations of them, etc. I am writing of writing, sympathetically, in advocacy, in the joint understandings that we have no dominion – they are not for our use – and that their rights to life, and to maximisation of the quality of that life, are much as we conceive our own. Writing enters a new place under such beliefs. In a sense it has to wrench itself away from itself. I want to write, with that new place in mind, of writing as a means by which we might approach that place.

But there are problems. Even the word *animal* itself is a problem. There is no such thing, Derrida says, as 'animal', let alone 'the Animal', unless the latter be taken as a conceptual construct, much as we might speak of 'the Republic', 'the Infinite', etc. 'Animal' and 'the Animal', Derrida says, are *lumpen* terms, a form of intellectual violence.[1] There are only 'cat', 'dog', 'giraffe', 'cicada', 'kangaroo', 'rat', 'ant', etc. If we speak of animals, then it should be of the specific animals to ~~which~~ whom we refer. The term *animal*, cum Heidegger, who devotes much of *Being and Time* to this project, is a concept, a construction, in order to define (and

---

1    As I have cited more than once, but the point bears repeating.

keep 'cat', 'dog', cicada', etc., from) the human, largely by listing what the 'human' is not.

It may therefore be that *animal*, in a piece such as this, can be used only under erasure. The author, too, is an animal. One of the first things we should do, when we contemplate *writing* (the) animal, and after acknowledging the intellectual violence involved, is acknowledge that we are *human* animals proposing to write about *non*-human animals.

There are challenges even here. Were we to follow Derrida's lead, for example, and, resisting the violence of 'animal', use instead the names of all the animals to ~~which~~ we refer, then when we wished to refer to that realm which we have in the past called 'animals', we would have to compile so vast a list we would in effect silence ourselves before the sheer weight of it. Yet non-human animals are voiceless without us. Silence ourselves we cannot. It is vital that we perform this work of writing animals. The 'intellectual violence' of imposing an umbrella term upon the vast array of creatures upon ~~which~~ we have hitherto imposed that term, moreover, extends to many of those creatures themselves, in that *cat, dog, rat, cicada* are also umbrella terms. There are many species of cat, of dog, of cicada, and, of and within these species, countless individuals.

Am I balking at the terms *animal, cat, dog*, etc., or am I balking at language itself, one of the central features of which, surely, is just such generalisation, and the way it makes the 'world' malleable, portable, articulable? But language is also the product of its culture, is deeply imbued with that culture, and is one of the principal, if not *the* principal, means of transmitting that culture. And the vast proportion of human cultures – the cultures of human animals – have since their inception been dependent upon an exploitation and suffering of non-human animals reinforced by assumptions of human dominion. The languages of these cultures are deeply imbued with these assumptions and a vast array of mechanisms that have been generated by and sustain these assumptions. From an animal(ist) perspective, they've become something of a hostile medium. But it is also, of course, a necessary and unavoidable medium. One of the aspects of writing non-human animals is that it involves one, inevitably, in a long and demanding process of undoing (dismantling, sidestepping, avoidance, creation of new pathways). One of the greatest challenges of this process is to ensure that it does not so disrupt one's communication that one confuses, alienates, or otherwise loses the interest of the very audience one wishes to address. The writer of animals walks a tightrope.

Non-human animals have been effaced or occluded in so many ways and at so many levels that there is little point in suggesting some place where this work of undoing begins, but for the writer it will inevitably involve a continual vigilance at the basic textual level. You might, for example, wish to write of the Red kangaroo as a creature of the great Australian plains. You might wish to speak of it as a species that has made remarkable adaptations to the landscapes of saltbush and red earth. You might talk of its speed, its endurance. But already you're in the territory of error and disrespect.

One of the simplest and most pervasive ways in which we distance and degrade non-human animals is by application of the third person singular neutral pronoun. Although the *Oxford English Dictionary* tells us that the word 'it' is used 'for animals', no creature is an *it*. A kangaroo is either male or female, a *him* or a *her*, and even that might be reductive. There are exceptions, but almost every animal, human or non-human, is a 'he' or a 'she', and we should attempt to specify wherever possible, and use 'he/she' or explain why it's not possible when we find that it is not. Likewise the term *species* must be used with caution, seen itself as a significant part of the species *barrier*. It may be appropriate in an overall taxonomy of earth's creatures, but how much do we need to allow its cold distancings – its emphasis on essential (*specific*) difference rather than familiarity, its shifting of attention from individual to *specimen* – into our *relations* with other creatures? We human animals are one of the tribes on earth (a particularly dangerous and destructive one, I think), Red kangaroos are another.

We might also give some thought to the term 'Red kangaroo' itself. Taxonomy as we know it can be distinctively patriarchal. Most individuals in the tribe of 'Red kangaroos' are in fact a bluey grey, but those are females (sometimes called 'blue flyers'). The tribe has been named according to the colour of its males. So too the tribes of the striking 'Blue Satin Bowerbird' and 'Golden Oriole', two thirds of which – the females and the juveniles, whether those juveniles be male or female – are distinctively *green*.

'He/she', 'his/her'; '*human* animal', '*non*-human animal'; the use of 'animal' itself under erasure, and the search for alternatives to it, and the substitution, where appropriate, of 'whom' for 'that' and 'which': in these and a great many other ways, writing animals can seem clunky, awkward, unwieldy. But most change is so for a time; it is the prejudice, abuse and disrespect that, a part of the culture for so long, have become naturalised.

Even our substitutions can leave much to be desired. The use of 'human animal' and 'non-human animal', for example, is itself brutal in its way, maintaining as much violence as it seeks to avoid. One of the principal problems with 'writing animals' is that we are drawing them – *without their knowledge or permission* – into this brutalising space of language in the first place, a space in which even the simplest word, unwatched, may in effect draw blood. But what else can we do? Most medicines are poisons if used unthinkingly. Language is no exception.

We are forced to be inventive; we are forced to find alternatives, take different routes. We should, for example, be sensitive to idiomatic uses of non-human animals ('killing two birds with one stone', 'the elephant in the room', 'behaving like a pig', 'the herd mentality', 'other fish to fry', 'sheepish', etc.: they litter the discourse in their thousands) and, unless self-consciously *undoing* them, avoid them wherever possible (indeed insist to ourselves that it *be* possible). We might speak of a certain poetic *nous* becoming involved, even in the writing of the most direct and emphatic prose.

We should also be particularly sensitive to, and prepared for, the fact that merely writing sympathetically about non-human animals is likely to draw criticism. Most particularly, we are likely to be accused of *anthropomorphism*. I should explain why I do not think we should be much bothered about this. To *anthropomorphise*, as I've argued elsewhere,[2] is to treat animals as if they were in some way human. To speak of them, for example, as if they were capable of reason, or could experience 'human' emotions, such as 'love', 'grief', 'boredom' or 'shame'. This is, apparently, a poor and un-objective way of conceiving. Science – for, although the proscription against anthropomorphism has spread widely through the humanities, it's from scientific quarters that the

---

2    See, for example, 'An Exoneration', above.

condemnation most often comes – would have us use instead such terms as 'pair bonding', 'separation anxiety', 'territorialism', 'dominance', etc.

I have several objections to this proscription, but I'll restrict myself to two. Firstly, the idea that we can view anything from some position of 'objectivity' above or beyond the human is itself naïve. As Nietzsche said (in *Human, All Too Human*, 1878, §9), we see all things through the human head and cannot remove that head. Merely to *think* is to anthropomorphise. Second, and more important, barbarity itself begins with the thought that we are so different from the creatures we live amongst that we cannot know or even hazard how they feel. This is not only a lie to ourselves, for in many cases in the experience of almost all of us, we *do* know how some non-human animal or another feels (*at home* the scientist *knows*, in many respects, how his/her dog [cat, bird] feels, and yet 'officially', in the laboratory – this phenomenon is known as *doubling*[3] – has no idea), but, since what is called anthropomorphism is central to what we call empathy, and empathy the key to compassion, in the denial of anthropomorphism is that repression of compassion which is fundamental to the abuse of animals that has always scarred this civilisation (and almost every other civilisation one can think of).

A principal concern of those who oppose anthropomorphosis is that, in the subject's extending themselves to the object in this way – in the human animal's allowing him/herself to make assumptions concerning the non-human animal – the 'objective' perspective of science is compromised. An interesting facet of this compromise, from an alternative, 'animalist' perspective, is that one cannot rule out that, just as the vehicle of a metaphor can indefinably but almost unavoidably influence the tenor it has been selected to elaborate, the link thus established conceptually in some measure becomes a conduit, and the object, infiltrated, in its own turn infiltrates the subject. In extending ourselves to the non-human animal, it may be that we are allowing,

---

3   A term I take from Lifton (1988), who outlines the psychological splitting which enabled medical staff in the death camps to deal with the work required of them. (Please note that, and some hideous experimentation upon non-human animals notwithstanding, I am *not* drawing any comparison between scientists and workers in the death camps.)

albeit in ways we might not know or be able to control, the non-human into ourselves.

All of which is *not* to say – and the tendency, by those who wield it as a critical weapon, to use the term as if *all* anthropomorphism were naïve notwithstanding – that there is not a 'naïve anthropomorphism', or in any way to support such naïveté. People will dress non-human animals in human clothes; Xerxes will have the waters of the Hellespont thrashed when a storm destroys his pontoon bridges, etc. The anthropomorphism I defend, and a great deal of the anthropomorphism criticised as such, is somewhat more sophisticated: an attempt – if I were to isolate one pre-eminent characteristic – to *know* non-human animals, not to efface them.

But there is knowing and knowing. One must attempt, as best one can, to know the creature of whom one writes (*'write about what you know'*), but that knowing must be against a tide of already-accumulated knowing some substantial part of which might be better described as obfuscation: an animal constructed in some part by actual experience, yes, in some part by examination, yes, but also by prejudice, use, mythology, superstition, etc. Are we to know the cow, or the cultural/textual cow? The rat, or the cultural/textual rat? The deer, or the cultural/textual deer? It becomes incumbent upon animal writer-advocates – those human animals who would offer voice to otherwise voiceless non-humans – to know as best they can the 'actual' non-humans for whom they speak.

Ideally this means physical encounter and observation, the more extensive the better. Physical contact, however, let alone the time to come to know and/or live with the non-human animals concerned, desirable as it may be, is not always possible: there is not always time – the suffering of non-human animals is dire, and their needs are urgent – and there is not always circumstance, the vast majority of us being urban dwellers, significantly distanced from the non-humans we might seek to aid. What *is* possible is to approach the texts and other resources that convey information about the various tribes of non-human animals with caution and scepticism, checking and double-checking the points upon which one, as a writer of animals, proposes to rely. One thing that has impressed me deeply about those creatures I have myself most sought to advocate for – sheep, kangaroos, pigs, ducks – is not how much is known of them, but how little, and how, over and again, a supposed 'fact' about them may not be that at all.

But now a digression, and a digression within that digression, by way of compound illustration.

Four sheep live with me. Let me, for ease of reference, call them Jack, Igor, Andrew and Kim (i.e. they are named, but not with these names). Even in introducing them there are problems. I will not say they are 'my' sheep, or that 'I have four sheep', since these imply ownership, and no animal should 'own' another. And when I say 'live with me' I imply a measure of choice, but what choice have they really had? They are sheep 'rescued' from the processes of industrial farming, and three of them, horns cauterised and tails cut off in infancy, bear visible scars of that process (the fourth, brought to us as a five-day-old lamb, bears different wounds). They are surrounded by fences and locked in at night for their protection: were they to get out onto the roads they might be killed, stolen, dragged back into the industrial process (slaughtered), and there are foxes and packs of wild dogs about. In effect, lovingly treated as they are, they are prisoners. Every four months or so, until we contrived a kinder, chemical-free method, they had to be dosed with an anthelmintic – a poison – to keep them free of worms and intestinal parasites, and at least once a year they have to be shorn. Sheep have been bred to grow as much wool as possible. To be left to carry a full coat through a hot Australian summer is a great cruelty, and while there are occasional escapees who manage to avoid being shorn, sometimes for years, their burden is oppressive and quality of life deeply compromised. *Not* shearing the sheep who live with me is not an option, and yet, even with the gentlest shearer, the process is undignified, uncomfortable and brutalising: there's little to no chance they'll get away completely wound-free.

I should say too that these four are all males. All castrated, two before they came to us and two, the lamb and a mature ram, afterwards, in the latter case a complicated and demeaning story, at least as far as this human animal is concerned. A point I finish upon because, well intentioned as it might be, merely writing about animals in this way, without their knowing or permission, is a kind of *further* imprisonment and castration. We should always bear in mind, however much it may seem as if we have no choice in the matter – that is, that we *must* speak

– that non-human animals neither give us permission to write about them, nor *authorise* what we say. The bruise (/shame) we bear for this is part of the territory: a legacy of aeons of abuse. Writing animals is an inherently ethical activity, but the ethical path is not always clear, and the ethics of human and non-human animals are not as often in accord as we might presume.

And now the digression within the digression, a short piece written for a different purpose, but serving this one even better:

> At the bottom, say, of a long stretch of boredom (but is it truly boredom, if motivated by curiosity, a desire to know?), self-imposed, watching sheep, forcing oneself to do so, to stay there, in the middle of the paddock, below the water-tank, resisting the pressure to *get on with* one's (my) active, understood, *human* life – a kind of meditation, yes, although even to see it thus is to distract – there is the possibility of seeing, noticing something one has not noticed before: Jack's sudden limping, could that be a pulled muscle? a stitch? Or the rhythm of their grazing, two snips upward and one back, like a barber, but is it the same for them all or do they have different techniques? And their rumination, their chewing, two of them clockwise, the others anti-clockwise: are there left-handed sheep and right-handed sheep? (Kim who was anti-clockwise yesterday is clockwise today); and the moods Igor wakes in, carries on through the day, do they come from something he's dreamt about, or might he have headaches, a tooth-ache? And Andrew, is it really true that, when you stroke his ears, his cheek, he has begun to purr like a cat (Jack's nose runs all winter, on the colder days, as does mine) …

It may seem strange that, in speaking of knowing and actual encounter, I speak in almost the same breath of boredom, of idleness. These are in fact issues – etiolations – of the gaze. The encounter, the gaze towards the animals, will seem so much an encounter with nothing: that is (a double encounter), it will seem to *give* us nothing, *produce* nothing (non-human animals will not *answer*), will seem *fruitless*, but paradoxically its fruit will be that Nothing which we have arguably built our civilisation as a wall against: that Nothing which now has to be the thing that teaches us, that dark, fig-like nothing which, opened, contains so many *seeds*.

So that (of course) it is not *nothing* at all. So simple, this knowing: you wouldn't think it encoded a revelation. Remembering – extending – the body, employing the (animal) self and that self's experience (one's own pulled muscle, one's own runny nose, one's own morning moods, one's own toothache), so that one *can* notice. Call it empathy. Assumption of a commonality of feeling of self and other; use of such an assumption to create (the illusion of) understanding of the other. An assumption that the other *is* like the self. All of which, for those inclined to do so, may be seen to loop us back to anthropomorphism, though I am here more concerned with the body, its use to us in approaching and understanding non-human animals, a vast and (bizarrely) largely untapped resource.

But this is a piece on writing and activism. Where is activism here?

ATTENDING AN ACTIVISTS FORUM in Melbourne several years ago, arriving late – we'd driven from Sydney – and finding no one at the registration desk, my wife and I set off in different directions to sample the parallel sessions. I wandered into one that proved to be a Skyped interview with an activist in Europe who had kindly got up very early to make the Australian connection. I remember little of what he said other than one very simple point that struck so strong a chord I have regarded it ever since as a truism. Neither the *numbers*, he said, nor the *horror*, of chickens suffering in a battery farm, of creatures killed each year for human consumption, of pigs slaughtered this week or that in an 'animal processing facility', carry the power, the mind-changing affect, of *story*.

It is widely and truly asserted that a great many of us, and a substantial proportion of those whose minds and hearts we seek to win to the support of non-human animals, suffer from compassion fatigue. This fatigue has various causes. It may be a result of having seen so much suffering and done so much to try to alleviate it that one has exhausted oneself, finds oneself run dry. It may be that, faced with the sheer numbers, extent and intensity of suffering, one has come to feel overwhelmed, powerless, wonders what, given such figures, such magnitude, any individual can do.

The question is not the fatigue itself but how, as writers, we might break through it. If something is going to do so it won't be numbers – numbers or, with a nod to Borges ('arguments convince nobody'[4]), arguments. The death of one chicken in a battery cage, one calf in a slaughterhouse, is an outrage. The death of five hundred is five hundred outrages. How are they to be felt? Can our feeling intensify so much? Exhaustion – the dissipation of the outrage – seems built into the attempt. A well-chosen, well-told *story*, on the other hand – a powerful *vignette*, an eye-opening *image* – can be a different matter. The most effective, and *affective*, focus, the most effective way to stimulate and maintain the outrage, is by imagining in and through the individual. (That tightrope: to avoid the pitfalls – the rocks and resistances – of the language, and at the same time use it as powerfully and incisively as one can.)

'As the mercury was hitting thirty-seven degrees that afternoon,' begins a recent piece on kangaroo rescue:

> the smell of her infected and decomposing body could be sensed forty metres away. The rescuers reached the site to find the kangaroo still breathing but unable to move. Skin and bones, badly-worn down teeth, pouch lacerated and severely infected by maggots (as was her mouth), death for this twenty-year-old lady was imminent.[5]

'Most people would have killed her,' the piece continues,

> and by doing so have acted in compliance with the Code of Practice for Injured, Sick and Orphaned Protected Fauna, issued by the NSW Office of Environment and Heritage ... With an emphasis on population health rather than individual health, the Code ... prescribes mandatory euthanasia for animals that have lost their reproductive capacity. Animals who exhibit signs of old age or whose ability for long-term independent food acquisition

---

4   A statement he attributes to Emerson (and others) but which I think he has generated himself.
5   T. Brooks-Pribac, H. Bergen and R. Mjadwesch, 'Not by Milk Alone: Attachment Relations and Wildlife Rehabilitation' (unpublished).

and processing is impaired, as in the case of the kangaroo above, subsequently named Thelma, should also be euthanized.

Luckily, this account concludes,

> the rescuers on site that hot midsummer day were Professor Steve Garlick and Dr Rosemary Austen, of the Possumwood rescue centre. They rehabilitate between two and three hundred sick and injured animals annually. Thanks to their care and expertise, Thelma recovered 'and lived out her days on a private property where her natural diet was supplemented with a high protein compound to prevent her from suffering malnourishment due to her worn down teeth.'

Story, and an individual non-human animal foregrounded – a gambit all the more significant given the way 'normative' narrative tends to hide or sequester such animals, restricting their access to and influence upon the author's wider world.

'By degrees the bracken thinned,' runs a very different passage,

> affording a view of a field that extended as far as Covehithe church. Beyond a low electric fence lay a herd of almost a hundred head of swine, on brown earth where meagre patches of camomile grew. I climbed over the wire and approached one of the ponderous, immobile, sleeping animals. As I bent towards it, it opened a small eye fringed with light lashes and gave me an enquiring look. I ran my hand across its dusty back, and it trembled at this unwonted touch; I stroked its snout and face, and chucked it in the hollow behind one ear, till at length it sighed like one enduring endless suffering. (66)

Arguably more can be said – and achieved – in a brief passage of finely seen, felt and written prose than in a volume of special pleading. The story need not be teased out: a powerful and incisive image is often so for the latent story it brings almost inevitably with it. *Imagination* – the ability to see what is not in front of you, the ability to see with eyes other than your own, the ability to conceive, and represent, things which you have not yourself experienced (and yet, paradoxically, an ability dependent upon our own experience) – is key, given that none of us *has* the eyes, or the experience, of the animal of whom we speak. The passage immediately above is from W.G. Sebald (*The Rings of Saturn*, 2002). A more recent piece comes from Melanie Rae Thon who, in a story entitled 'Galaxies Beyond Violet' (2017), writes about a bee:

> As she flies, foraging for nectar and pollen, the friction of wind through feathery hairs builds a static charge, her body electric. Above or below, the flower opens: infinite blue, worlds of yellow, a murmuration of white shimmering into thirteen thousand eight hundred lenses. She's blind to red, but sees a universe we can't know, galaxies beyond violet.
>
> *So lovingly she lands!* (117–18)

ANIMALS ARE A FRONTIER OF WRITING. How could it be otherwise, since they are in so many ways a frontier of (*human*) thought itself? And that frontier is not well mapped. Challenges emerge as one goes.

It is likely our first instinct, in writing of non-human animals, will be to approach them through the forms that are most familiar to us, but the mere presence of the non-human animal – of thought *to* the non-human animal – will pressure those forms. If, from one direction, we must be prepared to face the criticism of anthropomorphism, we must also be prepared, from unattuned readers (the very audience we seek to influence), to face the criticism that there is *too much* non-human animal in a piece. The animal *estranges*. Drawing non-human narratives towards the foreground, muting or displacing the normally-foregrounded human, our essays, stories or novels can come to seem texts turned inside-out. Our understandings of (say) the short story or the novel, and of the narrative arts that hold them together, are more deeply human than we might have imagined. Writers of *human* tales, for example, can rely upon – in fact deeply exploit – a broad consensus as to the key events of human lives, but to what extent are these comparable to the key events of *non*-human lives (what *are* the key events in the life of a magpie, a sheep, a rattlesnake?)? As we *follow* non-human animals, as we are *led* by them, as we place them, and not human animals, at the centre of our writing, they will pressure those deeply human forms (and their readers) and not be comfortable within them. In writing the animal we must be prepared to let go not only of ourselves, but to some extent even of the modes and logics in which we are used to expressing those selves, if only because it is no longer those selves we are most concerned to express.

Daunting challenges, it may be – the tightrope – but I do not see that they should put any writer off. Non-human animals need all the help they can get, and each writer coming to their aid will do so from their own particular vantage point. I state the case in its strongest form – what point in doing otherwise? – in the thought that people will take up as much or as little as they find useful or pertinent. It might be a comfort, for any writer contemplating the nexus with activism, to think that, when it comes to 'the animal', writing is of necessity *an activism in itself*, and that the obstacles, the rocks and resistances that occur in our path as we write non-human animals are also resistances, rocks in our thinking. As small as they might seem at first (*whom* instead of *which, he* or *she* instead of *it*), they are in fact the edges of – testament that we are entering – something vast. Although there can be no guarantee that we will take it on, we are on the verge of one of the greatest changes in human direction since long before anything we might have called the Enlightenment: to turn towards, or *back* to, non-human animals, a pivotal moment because it's been our turning away from them (and a turning that turning-away into use, consumption) that's at once enabled us to get as far as we have, and has doomed that going.

# Cull and Culture

*Although killing is often the first response in dealing with real or imaginary ecosystem challenges, killing is an intrinsically violent act, and as such it is ethically unacceptable. The act of killing also disregards the scientific consensus on nonhuman animals' sentience, which compels us to avoid considering nonhuman animals as mere numbers and objects, the latter being a familiar perception in conventional conservation. A paradigm shift in conservation is urgently needed – such that would recognise, admit and act upon, on the one hand, nonhuman animals' sentience and their complex emotional, psychological and social lives, and on the other hand human responsibility for many of the conservation problems we're facing today.*
   – Teya Brooks Pribac, 'Fact Sheet 2466'[1]

THE ETYMOLOGY OF CULL has nothing to do with that of *culture*, although in Australia you could be forgiven for thinking otherwise. Here, it can seem all too often that the idea of killing is almost synonymous with that of land or 'ecological' management. Not all that surprising, I suppose, when one considers how rapidly the application of inappropriate farming and grazing practices has damaged so much of

---

1    https://www.arcohab.org/factsheet.

the friable soil surface, and created such intense competition for what fodder the land can still produce, or that the encultured end point of these practices is almost always slaughter. When you are determined to raise cattle and sheep in a land so unsuited to them, a land where their hooves will effectively destroy so much of the productive capacity of the ground they walk upon, then *of course* the rabbits, the goats, the kangaroos, the wild horses, the wild camels, will be seen as enemies, competitors who must be eliminated.

It doesn't take a great deal of thought to see such a mind-set, bred of such misapplication and raising-to-kill, becoming entrenched, extending to other forms of agriculture, leading to the culling of kangaroos, flying foxes, cockatoos, galahs, koalas (now virtually extinct in many parts of the country), sharks, crocodiles, and whatever else either creates some irritation to the human animal or embarrasses it by stirring its conscience.

Go to the National Parks and Wildlife Service with a 'problem' with kangaroos on your property and it's likely that, for a few dollars, you'll be offered a licence to shoot as many as you require; go to the same directorate with a request to be allowed to *relocate* rather than kill those kangaroos, and you'll face such expenses, red tape and preconditions that all but the strongest and most determined will give up the idea entirely. A country which at times has seen itself a world leader in conservation and care for the environment would appear to have reached – but has it ever left it? – a point where the massacre of other living creatures is considered a standard conservation procedure. Conservation by killing. What the?

I'd rather begin this essay somewhere else. The old causeway, let's say, over the swamp at the bottom of our second acre, recently restored so that the four 'rescue' sheep we shelter don't get their hooves wet as they cross to the rich grass on the other side, nor damage the delicate ecosystem as they do so. What a pleasure it was to go down there one warm day not long ago and find that several of the tiny, endangered lizards for whom this hidden valley provides a haven were lying there, on the old railway sleepers, warming their chilly blood, and that the wild wood ducks had been using the sleepers as a perch and training place. We'd been wondering where this season's broods had got to, but it seems at least two of them have been there, learning to swim, out of sight, amongst the safety of the reeds. They can't fly yet, have to walk in, and the causeway – uncovered, repaired – has re-opened territory.

When I first *heard* the wild ducks splashing in the swamp (didn't *see*: the scrub was too thick) I thought they were kangaroos, but no. There *are* roos in the gully – and swamp wallabies, going by their scat – but in the seven years we've been here I've not seen any on this tiny, two-acre block. I mention them only because it's that time of year again. The government of the Australian Capital Territory, three hundred kilometres to the south, is embarking upon its annual slaughter of kangaroos. The 'target' this time is just over 4000. Last year (2018) it was around 3500. And we – and a lot of others – find ourselves heart-sore and impotent once more. The ACT government is fairly small, but it's a bureaucratic juggernaut nevertheless; from what I've read of the documents, shameful as they are, it will be almost impossible to stop the process or even change its direction. Forward estimates seem to see to that.

The official reason given for this annual cull is that it's in the interests of biodiversity, that the overpopulation of kangaroos in the area is depleting the native grasses (the 'biomass') and endangering, for example – depriving of habitat – such small reptiles as the grassland earless dragon and the striped legless lizard. Each year the ACT government settles upon a number of kangaroos to be slaughtered, to maintain what it deems an appropriate population. But of course the city of Canberra has been relentlessly expanding, taking over more and more of the kangaroos' habitats. The chief depletor of native grasses is without question the city itself. Even the ACT government admits that.[2]

One could speculate that the real reason for the killing of the kangaroos is that, in the face of this reduction of habitat, they have begun to urbanise – what other choice have they had? – and to enter the suburbs, graze in the city parks, become a 'threat' to cars and bicycles on the streets and avenues and parkways, that the frequency

---

2    'Is it fair', wrote Shane Rattenbury, MLA, the ACT government minister at that time responsible, 'that individual animals must pay the ultimate price for the effects of ever-encroaching human settlement?
       'No it's not fair. But it's the right thing to do.'
       – letter to *The Canberra Times*, 21 May 2015.

of accidents caused by cars hitting or trying to avoid kangaroos has increased; that they have become an urban problem; that humans can't bear the sight of kangaroos dead or dying on their roads, the thought of possible damage to their vehicles (and car insurance premiums going up) or of possible injury to themselves. And as usual the easiest and most economical solution is killing. Rather than line the major avenues with fences, say, and create underpasses at the most frequent kangaroo crossing-points, or reduce the speed-limit on such roads, or embark upon an expensive program – they say it's impossible but over and again they've been proven wrong – of relocating or de-sexing the animals. Better, after all, that thousands of kangaroos be slaughtered than a couple of humans die.

But of course, as already indicated, the official excuses are different. Conservation, ecological and environmental balance, biodiversity. In principle I have, for a very long time, been in great sympathy with conservation movements, but all is not easy in their relations with animal rights. One could almost say there's a schism between them, a divergence occasioned by the former's continual insistence upon culling or eradication of the exotic and/or supernumerary, at the heart of which is the deeply troublesome and paradoxical insistence that particular, human-animal-determined classes of *non*-human animals pay for the damage that *human* animals have done, and that the greatest cause of ecological damage and imbalance – the animal the culling of whom would make the greatest sense in terms of ecological repair – remains untouchable, to all intents and purposes out of the equation.

This is not to say that reparation, redress, attempts to balance, or minimalisation of further damage aren't very desirable things – they are paramount – but more and more, while it continues to ignore some of the large questions it raises, 'conservation' has come to look like an *aesthetic* rather than a defensible *ethic*, and a token gesture at that, a sideshow to distract us from the main event, the reality that the management of the ecosystem is in the hands of the top-order predator par excellence, and that, as any Australian government's budget will show, protection of wildlife and the environment are close to the bottom of its priorities.

I'm hardly the first to point out such things.[3] Indeed one could say there is – or has occurred – a schism in conservation itself, turning upon precisely this victimisation and scapegoating of non-human animals, and that from it has emerged the concept of 'compassionate' conservation. The conservation of which I have been speaking so critically in this piece – still the largest proportion of the conservation movement, and the source of most common understandings of conservation – is largely but not entirely that from which 'compassionate' conservation has separated itself.

Am I really calling, in my reference to 'the greatest cause of ecological damage and imbalance', for a mass culling of humans? Of course not. Good as such a thing might be for the 'environment', it would hardly be consistent (for example) with the principles of compassionate conservation itself. Humans – *homo omnicidens* – may have become, long ago, the top-order predators, but, predators or not, we are still animals amongst animals. I am, however, calling for some attention to the shaky foundations, unexamined assumptions, and paradoxes upon which certain current understandings of 'conservation' appear to rest.

What *is* it, for example, that we are conserving, or seeking to preserve or restore the balance of? What are its dimensions? How do we measure it? What are its borders, its edges? How do we define and locate them? Isn't (for example) every *macro*-ecosystem made up of *micro*-ecosystems? Can we be sure the assumptions we make about one (the macrosystem) – and the rules we base upon those assumptions – are appropriate to the others (the microsystems)? Do they have to be? And upon what point-in-time do we focus, to attain the model we wish, supposedly, to return our damaged/imbalanced ecosystem to? A point before human intrusion and damage? A point before *settler/ invader* human intrusion and damage? A point before the introduction of 'exotic' species? 1770 (the 'discovery' of the east coast of Australia by Captain James Cook)? 1788 (the first white 'settlement' of that coast)?[4]

---

3　Warnings of this schism, albeit within a North American context, were sounded as early as forty years ago. See, for example, Callicott (1980) or Sagoff (1984).

4　Again, I am not alone in pointing out such things. Others have expressed similar concerns – even, with varying degrees of success, attempted to propose solutions. See, for example, Thom van Dooren (2015), or Mathews (2012).

In 1788, it's been estimated, the kangaroo population of Australia was somewhere between one hundred and two hundred million,[5] a big gap, but imagine. And now the *estimated* kangaroo population[6] fluctuates between fifty million in a very good (i.e. kangaroo-friendly) year and twenty-five million in a bad one.

If the latter 1788 figure, the two hundred million, is correct (let's hypothesise) then the 'settlement' Australia has *quartered* the kangaroo population (in Tasmania quite literally *decimated* it). And yet we are told, by governments, farmers, conservationists, and the would-be moguls of the kangaroo meat industry alike, and in the face of an alarming lack of credible evidence, that kangaroos, at a quarter of their pre-1788 numbers, are in 'plague proportions', and should be culled (killed) relentlessly (the federal Department of Environment and Heritage would have us kill some four million kangaroos per year). And of course the kangaroo here is just one species among several we could be talking of. Once, they say, there was a koala in every tree. Now you'd be lucky to find a koala in every state, and their extinction is being predicted within decades.[7] So clearly it's not *1788* that conservationists are seeking to return us to. When, then, *is* it, and why and how has that choice been made?

Current conservation theory, if so rudimentary a notion could be called theory – better call it prejudice – would promote 'native' over 'exotic' species. While this has not as yet taken the institutional hold it has in New Zealand, which has set 2050 as the target year for the elimination (poisoning) of all exotic avian and mammal species, it nevertheless seems the prevailing rule-of-thumb of government and public conservation policy. But, plausible as such a policy sounds – who does not want to preserve Australia's unique native fauna and flora? – it is not without serious problems, contradictions and omissions. In Australia we human animals are ourselves exotic (the vast majority of

---

5 Auty (2005), 62.
6 The Commonwealth Government Department of Environment and Heritage releases a yearly estimate of overall kangaroo numbers on Australia, based upon state surveys and estimates. The technology of these surveys and the calculations based upon them are deeply controversial.
7 This essay was written between 2015 and 2018, that is, before the massive loss of koalas in the 2019/20 bushfires.

us, from one perspective; *all*, from another). Is there a deeper, more radical contradiction here? An hypocrisy?

Wasn't the system we would return the current ('invaded', 'contaminated') state to *itself* a changing rather than static thing? Can we be sure our insistence on repair and native/non-exotic purity isn't in part driven by guilt rather than sound reasoning? Isn't a system which relies upon human intervention to 'manage', 'protect' and 'guide' its return to a state *before* human intervention rather doomed and discordant from the beginning?

Why *must* an ecosystem exclude the human and exotic? Isn't the privileging of 'native' over 'exotic' a human concept in the first place, already an intervention? Can we ever be sure that the real engine driving our efforts at conservation and determining our understanding of an 'optimum' environment/ecosystem isn't our own survival? And how – to go back to my opening paragraph – can we ever be sure our ecological assumptions are integral, 'objective', *things on their own*, and not matters of particular/peculiar culture?

The insistence upon native over exotic – an especially *lethal* insistence, in that it not only insists upon the attempted eradication of a long list of 'introduced', non-native species (rabbits, foxes, feral cats, feral dogs, feral pigs, feral goats, rats, camels, horses, cane-toads, etc.), but in most cases does so through the use of poisons banned in the rest of the world, which in the process kill as many native animals as they do the non-natives they are deployed against – is particularly vulnerable to such critique.

Not only does the emphasis upon *biodiversity* weaken significantly in the face of the insistence that this be achieved through the *elimination* of exotic species, but the 'purification' thus sought, with its uncomfortable and as yet never satisfactorily dismissed resemblance to the kind of native-ground-based, *blut und boden* racial 'cleansings' of the Third Reich, amounts to a hollow and deeply hypocritical scapegoating of such staggering proportions that it's hard to see it as anything other than a symbolic smoke-screening, asking as it does feral cats, dogs, pigs, etc. to bear away the sins of the very exotic species – *humans* – who brought them here and, in their convenient exclusion of cattle and sheep (surely the most destructive exotics after humans), continue their sinning unabated.

What immense compromises have we already made, before we even begin to assert our ecological assumptions? There is little chance that any human-animal attempt to conserve anything is not already deeply *conditional*. We may have a will to conserve and restore, but it's unlikely we'll be prepared to relinquish many (if any) of our comforts in so doing, let alone leave or move aside or give back much of what we've already taken.

Here in Australia our understandings of conservation seem to be determined as much by a panic button as by any deeply thought, integrated and consistent ethic. Let some researcher determine that some species of bird or reptile or small marsupial is on the verge of extinction and we'll leap into action, placing fences around it, killing its predators, 'rescuing' its remaining members by taking them, away from their existing habitats, to research/breeding facilities, in such a manner as has led some to refer to 'zoo' Australia, which is probably not so bad a comparison. Zoos are full of cages that separate one species from another, always with at least one side 'open' to the viewing public. A museum mentality. Conservation by incarceration.

Is that what conservation's ultimately about? The human gaze? Not entirely, of course, but clearly it's a significant, even shaping component. Imagine it, this 'zoo' Australia: at the best it's going to be small areas carefully fenced/screened off from the surrounding countryside – the surrounding *eco*system – many of them 'preserving' one particular threatened species or another, and each of them increasingly vulnerable to human population growth, reallocations of land and resources, changes of political will, economic tightening (bushfires, floods, viruses), etc. A frightening idea in many ways but perhaps most in that, when, with all its illusions, it takes hold – when people accept that it *is* an option, an insurance, as there are increasing signs that they are doing – they will, thinking its future secured, all the more rapidly relinquish the wildlife – the lives-at-wild – we still have.

This must sound as if some of the fundamental principles and rationales of conservation have passed me by. I'm not so sure of that. A 'feral' cat, it's said, will kill up to four 'native' animals a night, which is to say that, even if we dramatically reduce, say by ninety per cent, the (doubtlessly exaggerated) claim made not long ago by the Federal Environment Minister that there are twenty million feral cats in

Australia,[8] the feral cat population will destroy several million other creatures in a twenty-four hour period (around two and a half billion native animals per year, according to his calculation), endangering the survival of many native species. This is a terrible situation by any account, and presents an appalling dilemma.[9] But is the only solution, to this or any other perceived over-population, eradication? Is killing the only thing we can think to do? A widespread carpeting of the land with poison? Of course not, but the alternatives – containment, capture and de-sexing, etc. – are, to most people's thinking, impossibly impractical, by which is meant – let us, in this 'clever' country, understand this very clearly – nothing more than *uneconomical*.

But whence this term 'uneconomical'? Isn't it only to say we have other things which, collectively, we'd rather spend our money and effort on? Designer jeans, say, and cars, nice wines, the latest trend in café brunches, a new fleet of submarines,[10] a further tax deduction for the already rich, another immensely destructive incursion, in support of one or another of our allies, upon some other country's sovereign territory. We may *want* conservation, but how much are we prepared to invest in it, to enable us to do it ethically? Not very much at all, it seems. Arguably almost nothing.[11] The first and greatest violence against non-human animals and the environments in which they live is *budgetary*.

In the hidden valley in which I live – that is, I call it a hidden valley, though in truth there is nothing very hidden about it – holly is a problem. Indeed the local council has determined that it's a weed, a pest, along with agapanthus, blackberry, alligator weed, boneseed, and some hundred or so other exotic plants introduced by and since white

---

8    https://ab.co/3mtheW6, but see Legge, Murphy, McGregor et al. (2017), who set the figure between 2.1 and 6.3 million.
9    See van Dooren (2015), 22.
10   Defence department officials in November 2019 estimated that Australia's new fleet of attack submarines would cost $225 billion to build and maintain.
11   A survey published by the Australian Broadcasting Corporation in October 2019 found that, of the 54,000 Australians responding, 21 per cent were not prepared to contribute anything towards fighting climate change, 18 per cent were prepared to spend $100 per annum or less, and 13 per cent ticked the 'Don't know' option.

settlement. And the council poisoners – who've described themselves to me, very emphatically, as conservationists (it's even painted on their truck ...) – have numerous times, as they've sprayed a not-to-be-named and highly controversial weed-killer on the ivy and broom and blackberry that have come to dominate the copse at the bottom of the dead-end road beside us, asked me to eradicate it (and the ivy, and the blackberry, and the broom) from the property.

I don't really have to, as it happens: the sheep (also unwanted exotics in the area) who live with us tend to take care of much of that, being lovers of broom, blackberries and blackberry leaves (and roses, and camellias ...). But, to come back to the holly – and to say nothing of the creatures, native and exotic (the birds, the tiny, endangered lizards) that the council's famous weed-killer (widely dubbed 'Agent Orange Two') also kills – I was, on a misty afternoon a few Christmases ago, taken aback, indeed deeply enthralled, by the huge colony of spiders – *native* spiders – that one particular holly plant was sustaining, so much so that it seemed, almost, festooned with artificial snow. Where would *this* – surely an ecosystem of its own – fit in the council poisoner's ecosystem? But what ecosystem would exclude it?

So too – another observation – we are told, and I readily believe, that we have named and 'identified' only a small proportion of the mushrooms that grow on the lawns and in the fields and gardens and forests around us, only a small portion of the many strange and wonderful moths and other insects that (say) crowd the dark-side of my kitchen window on a warm summer night (alas less and less of them each year: the Realm of the Insects, in the face of climate change, is shrinking at an astonishing pace). Favouring big creatures over small, natives over exotics, the more visible over the less visible, we seem, conservationists and animal rights activists alike (and these – conservationists and animal rights activists – *do* seem to be far more subtle dancers than the rest), to be very dangerous, bumbling, thick-witted creatures indeed in our understanding of ecosystems, about as in touch with the realities of our own being – sensitive, say, to the soil we walk upon, and all the small lives there – as carnival stilt-walkers.

SPECIESISM. *Distance.* This morning I woke thinking of the ACT kangaroos, and of the way they are 'managed', unaware at first of the anger that was rising in me, from wherever it is I characteristically try to banish it to (in the interests of the 'clarity' and 'elegance' that should mark a good essay, for example), and of how it was rapidly becoming a cold fury. At myself, in part, *for* that banishing.

These fellow creatures, these mammals like ourselves, these *individuals*, with every bit as much right to *be* on earth, to *live undisturbed* on earth, as we claim for ourselves. And those who shoot them – who, according to instructions they've received from the desks of bureaucrats, select which ones are to be permitted to live and which are to die, who align within their sights those who've been selected, using the head or the chest as target, and who pull the trigger, who feel a sense of relief and achievement when the creature drops – or those bureaucrats who instruct them: how much do they *know* the creatures they kill, *as* individuals? Have they walked among them? Well, yes, some of them have, as they've counted them – the ACT government has given ample explanations of how populations to be culled are *counted* ('the five methods'), of how 'target' (yes, 'target') population densities are determined – but that has not been for, in fact has carefully avoided, *contact.*

Have they sat amongst them? Have they spent enough time and observation to be able to tell one mature doe from another, one mature buck from his brother? Do they see any one of them as an individual, a being of existential intensity, alive in every fibre, with a character, desires, griefs, resentments, secrets, favourite foods, favourite places, parents, grandparents, children, grandchildren, good, better, best friends, brothers, sisters, uncles, aunts?

I imagine the answer is no; I imagine the answer is that they cannot be 'managed' if the relationship is 'personal' in this way – not that there is any intention, if and when they use the term, to impute, imply or concede that the kangaroo can *be* a *person.*

And what of the kangaroos themselves? What contact have they had with humans? They may have been approached, occasionally, by some – a child, say, or a tourist, a bushwalker, a naturalist even – but my guess is it's never, on the human side, been with the patience to sit a whole hour, a whole afternoon watching in silence, thinking, recognising: one living, intensely-feeling *being* being-with another. If there's been contact, before the shot that kills them, and other than with those who are counting them in preparation, then it's as likely as not to be shot in a different manner, as in *darted* so that, temporarily narcotised, they can be investigated, manipulated, by scientists, naturalists, ecologists, wildlife managers. And then, of course, the shot, or series of shots, and the being, the existential reality, of twenty, or forty, or fifty-five per cent of them – whatever proportion has been determined in order to arrive at the target population – will be removed, will cruelly and abruptly cease. At night, in circumstances of great secrecy, to be buried in unmarked mass graves in the forest (as I have said elsewhere[12]).

I am not speaking of our right to do this. We *have* no right to do this. To turn to the life of another being and snuff it out. It is bad enough that we deprive them of homelands, confine them in special, fenced districts ('Kangaroo Management Units'), herd and manipulate them at will. If anything I think I am approaching, if only for myself, a sudden new understanding of the species barrier – that is, those measures we take to maintain our perception of them *as species*, and *specimens* thereof, rather than as individuals, *needing* the broader classification to keep them at 'manageable' distance. Not so much that barrier between species that we might possibly seek to cross or break down, as the use of the very notion of species itself, as a means of distinguishing one group of beings from another, as a barrier in the first place: *language* as a fence, *concept* as a fence. A part of the viewer's weaponry. The human gaze again.

WHAT IS AN ECOSYSTEM? How big *is* it? Where *are* its borders? One might justify the concerns for preservation of an endangered species, on the one hand, and the elevation of *native* over *exotic* on the other, with reference to the desirability of biodiversity and particular ecological balance respectively, but while each of these recourses has a certain logic, and in an undamaged and uncompromised environment could almost be considered principal, the environment, just about

---

12 'Roogate' (2016), re-posted by the Australasian Animal Studies Association at https://bit.ly/33pRYaj.

wherever we look, is far from undamaged and uncompromised and such truths, claims or imperatives as those arguments (biodiversity and ecology) possess have to do battle with so many others – struggle, as it were, within an *economy* or *ecology* of likewise-relative truths and imperatives[13] – that we should perhaps consider that we are confronting a kind of paradigm shift in our valuations, or at least now find ourselves on a planet that is demanding one.

Exactly. And that we rethink – ask again – just what we mean by an ecosystem, and the extent to which our received understandings of it will increasingly pale and fail in the face of the challenges of continuing human predation. Don't we mean, by an 'ecosystem', something *out there*, that does not include us and our cities, our irreparable and irrepressible globalism, etc.? Too often our sense of 'ecosystem' is something with a conceptual, and sometimes very real, fence around it, a pure (or purified) enclave surrounded by and struggling to keep at bay our own immense corruption, pollution and contamination.

*Within* this conceptually sequestered ecosystem, paradoxically, there are (supposedly) no borders – that is the very idea of an ecosystem, as a thing of balance, self-regulating – or, at least, only 'natural' borders, created by terrain, microclimate, soil composition, etc. And yet, of course (the paradox), the ecosystem, *that* kind of ecosystem, can only exist because of the borders we place around it, needs our regulation – albeit regulation of ourselves – in order to 'self'-regulate. But what of the systems – the 'ecology' – of which such ecosystems are a part?

I am a lover of the Australian bush, of Australian flora, Australian fauna, and wish to minimise, in any and every way feasible, the harm that humans do to it, but that word 'feasible' is a crucial and troubling one. Isn't an ecosystem that is in effect an island, a prison, a zoo, a realm frozen in an imagined time-before-us, a kind of paradox, a sort of contradiction in terms? To what extent is it sustainable, let alone realistic, compassionate?

---

13 Again, and although I have come to it independently, not an unshared position. Van Dooren (2015), for example, articulating a similar position, directs us to Haraway's idea of 'situated knowledges'. See Haraway (1991), 198.

This is not to say – although I'd clearly wish to insert compassion into our equation – that we cannot continue to 'conserve', indeed any alternative is unthinkable, but we have to be aware, as we do so, and if only as a prelude to attempts to strengthen them, of the fragility of the concepts we've been bringing to the process, on the one hand, and on the other of the vast, unpredictable and uncompromising Ecosystem we're trying to hold back from the ecosystems we're attempting to conserve. Although it may seem outrageous to suggest it, it may be, for example, that our own survival as a species is neither likely nor, as far as the planet is concerned, desirable. The environment and the shifting climate will determine – will make their 'choices' as to – which species live through their crises; it is ultimately not going to be for us to do so. It is, after all, an old saw of human management that, when looking for someone to repair something, one should not trust the person (or species) who created the damage in the first place. But of course, as a race, we blunder on.

As to the killing. That is a huge tide to turn back. It comes – a clumsy and lethal default – from a dark and shameful place in the national psyche and we must acknowledge that place and attempt some healing and reparation. The rest, and although this is a huge matter, is no more than a matter of will. This is the country that devised and managed to complete the Snowy Mountains Scheme, in which (for better or worse) two major rivers carrying the vast snow-melt to the sea were, by an extraordinary feat of engineering, *turned around* so that that snow-melt went westward to the Murrumbidgee Irrigation Area, turning a vast tract of arid land into orchard and generating, as it did so, much of the electricity requirements of Sydney and Melbourne for decades to come.

If we wished to do so, we would be able to find alternatives to our bizarre, deeply shameful, and deeply *lazy* killing-for-conservation. And, if we wished to do so, we would prioritise – we would demand – the requisite funding. It is up to us to turn ourselves around.

# Lisika

PEOPLE TALK ABOUT THE AREA as swamp country – there are many rivers and streams running through it – and sometimes refer to the locals as frogs, but in fact Hope is on a mountaintop, with verdant fields and stunning views of the alps northward. The sanctuary, of about eight acres, is home to six horses and a donkey, thirteen goats, a bull, four sheep and seven pigs, four of whom are still well under twelve months old. There are also three dogs, some hens, some rabbits, and at least three cats. I could have counted more carefully but we were busy.

We'd been called in by a mutual friend who herself has a micro-sanctuary (a backyard full of non-human refugees in enclosures made of pallets and old packing crates), and who is often asked to find homes for abused or abandoned farm animals. She 'rescued' a sow six months ago but had no room for her herself and so houses her at Hope and pays costs.

Soon after this sow arrived, Dora and Alin, the couple who run the refuge, realised – such discoveries are disasters for any small sanctuary – that she was pregnant, having been housed with other pigs, amongst them a boar, in the place she'd just come from. Hence the four piglets mentioned above. There had been nine in the litter. One had been still-born. Four had died soon after. Two of those had been badly deformed.

And this sow, Lisika, was now sick, hadn't eaten for two weeks – had been refusing food – and had just been taken to the vet. He'd rehydrated her with an intravenous drip and instructed she be fed small doses of food and water, by syringe, every half hour for the next few days. Dora and Alin were stretched to the limit – they always are – and had just been let down by a couple of volunteers, so couldn't take on this new task themselves. Nor could our friend whose micro-sanctuary is thirty minutes away and who has her own hands more than full. We found ourselves volunteered.

In the process of examining Lisika, I should add, the vet had found she was pregnant again. Four to six piglets, as far as he could tell. This was disaster redoubled. Any animals born on a sanctuary take up space and resources from other animals desperately in need of them. It's therefore common practice that males coming into a sanctuary, if they're to be with females of their own kind, are castrated as a condition of entry: one of the many painful ethical dilemmas of rescue. So how this second pregnancy had happened is anybody's guess. A wild pig may have come in from the forest somehow, or (more likely) Dora and Alin had waited just a tad too long before having the male piglets castrated, and Lisika had become pregnant to one of her own sons.

So. We are there. Or rather here, T. and I, for the weekend, to take shifts trying to get Lisika to accept food, or at the very least water, and perhaps do some other jobs about the place if we have time, to relieve the general pressure. Lisika – we are taken almost immediately to meet her – is lying in a darkened enclosure at the back of the barn, the rest of which opens onto a large yard which in its turn opens onto an even larger embankment below the house, reserved for the goats, sheep and horses. Several times, coming from the house or car, we have to weave our way between the horses or the goats to get to and from the enclosure.

We arrive at noon on Saturday and from then on – except for a period from midnight Saturday until about 9 am Sunday – Lisika is almost constantly attended. On our part it's T. who does the real work. With a bad leg, bad knees, bad sense of balance, there's little I can do but sit, watch and think, keep T. company, pass her pieces of cloth, medi-wipes, or her water-bottle to sip, or prepare syringes of water or pureed food (bread, banana, a touch of molasses) for Lisika.

The first thing T. does when she gets there is to lie down on the straw behind Lisika and put her arm over her, hold her gently an hour or more, trying to adjust her bodily rhythms to Lisika's, and Lisika's to her own. Body rhythms, rhythms of blood, rhythms of breath. From time to time she reaches over for a syringe of water or food and tries to encourage Lisika to accept it. At first it's much as it has been, apparently, since she came back from the vet twenty or so hours before – Lisika gritting her teeth, allowing nothing through – but after an hour of T.'s holding and stroking her, my wordlessly watching, as motionless as possible, and everyone else banished, Lisika actually opens her mouth and accepts some water and a little food. It's not much – 20 ml of each – but something, a beginning. Alin, when he comes down, jokes that they've found a pig whisperer.

From time to time one or another of the horses comes to look over the top of the enclosure gate. Most often it's a large and beautiful dark-brown mare, who seems to watch with real concern, even pushes my hand away when I try to stroke her, as if to say *I'm not here for that*.

In my camp-chair on the side, watching for hours on end, waiting for T.'s requests and instructions, I have a lot of time to think. Much of it is about what Lisika herself might be feeling and thinking. I try to remember a time, long ago, when I too was very ill, how it felt. At one point somebody – I can't remember whether it's Alin or Dora – asks 'Do you think she is *present*?' At first I think the question absurd, a matter of the consciousness we do or don't allow to animals. But I don't know. I should ask what they mean. I think of Derrida's writing of the abyss that he says he finds, and fears, behind the eyes of animals. When you place an abyss between yourself and non-human animals – conceive it, imagine it – you do all sorts of things, all of them negative. We humans are animals too; the abyss, it seems to me, is part of our attempt to deny this.

I don't mean to say I know what Lisika is thinking. I have no idea what is going on in her mind, and neither of us has any means of 'talking' to the other. But things happen nevertheless. At one point – I am up in the house, but T. sees this – Lisika *cries*, which is to say is seen, momentarily, to have tears running (T. says *gushing*) from her eyes. I've looked this up since and of course there are farmers and scientists aplenty to say that such things, in pigs, are merely emanations of fluid from the tear-ducts and no indication of emotion, but by this point, having watched her all afternoon, it's clear to me she's either experienced a severe wave of pain, or that a particularly sad thought has crossed her mind. Later I wonder if this thought, if that's what it was, might have been the realisation that all of the piglets within her were dead, or perhaps that the last of them had just died, but I'll come to that.

I also wonder – there is a lot of time to wonder – what part *refusal* might have in all this. I've looked up 'sows not eating' (etc.) and the reasons I've found are all to do with physical malfunction; something is wrong, as it were, with the *machinery* of the sow. There's no mention of *will*, or that the sow might *choose* not to eat, whereas when it comes to humans it's generally recognised that failure to eat can be deliberate: an infant's (or an elderly person's) only means of asserting some *control*, for example, or resisting someone else's.

Had Lisika decided to starve her unborn litter, to abort and have done with it? Cruel, yes, or desperate, but it might have come to that. Why should it be that only *human* animals can have crises of this kind? And of course she might have had other reasons. Julia Kristeva writes, in her essay on abjection, of the child's refusal of food as an indication of their having *swallowed their parents too soon*, something that has lodged firmly in my mind because for the first nearly twenty years of my life I refused to eat green vegetables (or any other colour, for that matter, except potatoes).

Pigs are very sensitive to being interfered with (to being *controlled*), and, early breakthrough as we might have had, Lisika, by mid-evening, is having no more of it. She is exhausted, as are we all. We leave her to sleep quietly. T. and Dora check in on her before going to bed, then she and I have a very late glass of wine under the stars before sleeping, by the barn, in a caravan full of the smell of old cigarettes. I get up at eight the next morning – it's already hot – to search for coffee, and check in on Lisika on the way. She's covered in flies and is lying in the product of her diarrhoea, a very sad sight. For the first moments I think she hasn't made it through the night, wonder if I am looking at a corpse in the very first stages of dissolution, but then she weakly, and vainly, flaps an ear against the flies. I'd get down and clean up her mess myself, just as T. has had to do a dozen times through the Saturday, but with my knees and balance I'd probably just fall into it, exacerbating the problem. Does Lisika know I'm there, watching? Is she *present*? I can't tell. The dark-brown mare is watching too, has probably been checking on her right through the night.

I get the coffee. T. wakes, takes hers into the enclosure and begins cleaning Lisika up. She tells me, when I come back a few minutes later with some bread, that there's something, a bit of a sac, protruding from Lisika's vagina. Birth has begun. The things one doesn't know. I'd had no idea, for example, that each piglet has its own amniotic sac. Dora comes down, and she and T. begin stroking, encouraging.

The first piglet is born, dead, and clearly several weeks premature, at about 10 am, and we place it – no piglet is an 'it', but I didn't register if this piglet was male or female – in a white plastic bucket, a very sad but also strangely beautiful sight. The perfection of the mouth, the tiny hooves, the skin, shining, purple-hued. And so it goes, on through the day. Three have been born by the time a second vet – not the town one – arrives at 1.30 pm. All dead. He gives Lisika an injection of oxytocin, to encourage the birth, and stays until 3.30, refuses to accept any payment. Another is born – also dead – while he is there.

After he's gone T. and I sit with Lisika another couple of hours (I sit, T. lies), though by this time she's exhausted and we think it better to let her rest than to urge more contractions. Dora comes down at six to watch while we go up and eat some of the dinner Alin's prepared, but soon she is calling for T., and another five, all dead, are born in a rush between 6.10 and 7.30. So much for a litter of four or six. We are now thinking – T. and Dora have taught themselves to feel them – that there is at least one, and perhaps two more. When T. comes back up to the house she says Lisika has just struggled to her feet – she hasn't stood the whole time so far – and then vomited. She also says that, a little after the birth of the ninth, tears had again gushed from Lisika's eyes.

We divide the night into shifts. T. and I will watch until 1 am, and then Alenka, Alin and Dora's teenage daughter, will watch through until dawn. When I go back down at eight, a little after T., Lisika has moved towards the chair I've been sitting in and is lying on her belly for the first time since we arrived. To me she seems a lot better, although soon enough, Dora gone, she rolls onto her side once more, and some strong contractions indicate that she's working to deliver another. Two hours then, of stroking, whispering (*push ..., push ...*), cleaning up messes, resting, trying to give her water to keep her hydrated, before the tenth is born, also dead, at 10.10, and then, after an hour's rest, the most extraordinary hour and a half of concerted exertion, T. having found a way to stimulate contractions, and Lisika working with her. There's no verbal language but communication nevertheless, and *effort*, on Lisika's part, such as we've not seen all day (though without question she's been trying), and the brown mare is again there, looking on.

It seems to me, watching, keeping the flies off Lisika's face, handing T. whatever she needs, that – and I'd rather not use the terms 'human' and 'pig' here – I am witnessing T. learning some very real things about Lisika, and Lisika learning some very real things about T. The space between them – to come back to Derrida's psycho-geography – might have remained abyssal in some ways but there are now rope bridges across it.[1] I feel, too – it seems almost that I can *see* it – that somewhere in this process Lisika has turned a corner, has come back to life, as it were, decided *for* it. We might simply say, if we wish to deny this – take away the possibility of volition – that she's begun to feel better and is responding, physically, accordingly. But that, too, would not be an un-human thing, and who is to say?

Where does this leave us, with regard to the abyss? I find these days I'm increasingly losing patience with hagglings over semantics in the absence of direct physical experience. Perhaps we force ourselves to *terms* too much – abstractions, concepts, and the finding of words for them – and they begin to encage and limit us, make us think they're the only options, the only way. One of the many things that come to mind as I watch (it seems my job, if anything, *is* to watch) is Bentham's 'The question is not, Can they reason? nor, Can they talk? but, *Can they suffer?*' Grateful as I am for his having stated the obvious in this manner, it seems also suddenly absurd, given the suffering that I *am* witnessing, and all the thinking going on within that presence and that

---

1    The human bubble. How much is illusion? I have written only a few paragraphs earlier that pigs 'are very sensitive to being interfered with', yet what was this but massive interference? Yes, in the interest of saving Lisika's life, but interference nevertheless. I write too about 'keeping the flies off her face'. What stays with me years later is the look in her eyes as I did so, which at the time I took as a sort of gratitude, but which seems to me now to have just as likely been exhausted contempt. More humans, interfering even with her death.

suffering, that this should have been deemed – *still* be deemed – a *question* at all. I *see* it, *there*, in front of me: how then can I ask if it *exists*?

Perhaps it's not a matter of whether there's an abyss or not, but of what force we need to give to it, if we need to give it any force at all. We should focus on bridges instead. The problem, of course, is that bridges – the building of them – are so demanding, so visceral, require us to *work*, face us so much with our own incapacity, steep us so much in the abject (for yes, that too is a part of it here: those ten small bodies now crammed into the plastic bucket; those six, seven, eight dozen cloths, sheets, bits of old tea- or bath-towel that we go through just cleaning Lisika as the process drags on), that there is almost a *letting go of oneself* required to even enter their territory (the animals', the bridges'). And whose abyss anyway? More than once, as I leave the enclosure, or just stand and go to the gate for a fresh sight or some fresh, horse-dung-scented air, the white goats are there and turn to me as if for news. I wish I could tell them. Perhaps, as they look at me, wondering what we've been doing, what's going on behind me, they feel that it is *I* who is the abyss. (For that, surely, is the other side of it, that the creature who returns our stare sees the abyss in *us*, an abyss from which so much violence emerges.) Which is not to say that, if I could spend enough time with them, I might not learn how to set them at ease. A year's observation, of *being with*, perhaps. More likely ten.

Another thing that comes to mind as I sit there – an old thought stalled upon a couple of years ago, abandoned as whimsical, but that now, in this space, seems to clarify: that the mind of the human animal (mine, here, now), steeped in a space with a non-human animal, can perhaps have some confidence that the thought that comes to it may be thought which, if only in some tiny measure, reflects or overlaps the thought that comes to the *non*-human animal there. If the things in *this* place, I find myself thinking – the straw, the pain, the processes (of birth, of feeding), the people and other animals (T., Dora, the mare, Lisika, myself, the goats), the clean cloths and the soiled ones, the dead piglets in the lidded bucket, the larger bucket of water, the smells, the light, the amazing, tiered cobwebs about the wooden beams above us – produce certain thoughts in me, certain *emotions* if you like, then surely the same combination of things, and other things here of which

I may be only unconsciously aware, might be producing or inflecting thoughts in Lisika also.

Mightn't it be, given this core of common material *in this environment*, that some part of what I am thinking might correspond to some part of what *she* is thinking (although *thinking* itself seems a constricting and inadequate term here: can I say *being*?), even if it is only regarding the position of a fly on her eyelid that she wishes I would brush away? Could it be that the more I am able to *be*, in this space, the more of what might be in her mind might in some way be in mine also?

I don't mean elaborate thought necessarily, and certainly don't mean in as much as any of that thought concerns things – the content of memory and experience – that neither of us can ever know about the other, but overall a possible *commonality of being in that space* that could mean that, instead of focusing upon the *im*possibilities between us, I might let my mind drift, open, accept, trust what comes into it. I don't mean to make any huge claims for this; perhaps only that the more we can occupy and open ourselves to the same spaces together the less we might be said to find ourselves in something like that thing which Derrida has called an abyss.

Which isn't to say – since any space is also a *confluence* – that this particular space might not have some power of its own. In some ways I've only just touched the surface. Dora is naked, for one thing. T. had forewarned me (only moments before we arrived) that Dora and Alin like to work naked – that they are, in essence, nudists. Not a problem. So long as I'm not expected to do the same. But the sight of Dora's large, strong body with her huge, pendulous breasts, moving about in the half-light of the stall, or sitting in a corner quietly breastfeeding her two-year-old daughter, naked *with* Lisika, reminds me so powerfully of my own mother, dead now forty-six years, that I can't keep her, my mother, out of my mind, nor, at the base of her so-similar belly, the deep scar from the hysterectomy, conducted in the process of my own birth by caesarean, that in effect meant that there were no more children after me and that contributed, I have often thought, to the deep current of sadness that led to my mother's early death.

How too, as she helps Lisika express each of these dead babies, counts to nine, hopes so much – she says this (and her eyes well with tears) later – that the tenth will somehow, miraculously, be alive, could

T. not have so much more powerfully in mind the fact that she herself had been the one in ten, the sole live birth among nine miscarriages? If there are resonances of this kind for us, might there not be also for Lisika, or the brown mare? I don't know. The world is a dark poem. One just doesn't know.

We go to bed in the musty caravan at one thirty, reluctantly and in two minds despite our exhaustion (mine nothing to T.'s, I suspect). The eleventh piglet has, with all Lisika's contractions, ascended from the base of her belly into the birth canal, and T., sitting behind her, stroking down, has now and again been able to glimpse the 'thread', as she calls it, that has so far been the first part of each amniotic sac to become visible – she has even, once or twice, thought she glimpsed the tip of the sac itself – but it's always slipped back, and we're not too confident that Alenka, alone, will be able to continue to encourage the birth. So near and yet so far. Hopefully, with more rest, Lisika will manage to get this last poor piglet and the placenta out by herself. If not, there is now an appointment with the vet in town at 9 am, and he might be able to do it for her.

We wake at 8.30 to sounds of heavy lifting and two or three robust squeals which, while painful to hear, also indicate that Lisika is still very much alive. Dora, clothed, comes to say goodbye, tells us that there's been no eleventh birth, no production of the placenta overnight, but that Lisika, when they entered, had stood, and had a long drink of water.

We pack, have coffee, leave. No choice about it. Much later in the day Dora calls to say the vet has managed to remove the eleventh piglet but that there's one more still to come. He'll keep Lisika in overnight to see if she can deliver this twelfth and the placenta herself. If not he's not sure what to do. She's too weak for a caesarean. But having spent the next night there, there is again nothing. The prognosis isn't good. It sounds to us as if Dora and Alin are being given the option of leaving Lisika at the vet's, or taking her home to die there. They choose to take her home and that seems like the right decision. Can a sow survive with a dead piglet and the placenta still inside her? We just don't know. From what I've been able to determine such a thing need not be fatal, but there are almost certain to be some hard problems to overcome. Puerperal fever, excessive bleeding.

Five days later we're still getting daily reports. On Wednesday, still on a drip, Lisika is up and moving about, and, Dora says, with pleasure in the irony, 'eating like a pig'. Thursday she isn't so well, but that might be because another pig has arrived, from Croatia (how did they smuggle him across the border?), and Dora and Alin, busy building a new enclosure, haven't been able to pay Lisika so much attention.

I keep remembering a story I read as a kid, in *Ripley's Believe it or Not*, about the mother of some pharaoh or another who, when her mummy was x-rayed by archaeologists, was found to have a dead foetus in her womb. They determined that she died well into her sixties, and that the foetus had been inside her since her twenties. Every year or so there is another such thing reported: an eighty-four-year-old woman from Brazil who's had an infant inside her for forty-four years, an eighty-two-year-old woman from Colombia who's carried a dead foetus in her womb for forty; a ninety-two-year-old Chinese woman who's carried an unborn infant for sixty. Apparently the body covers these infants with calcium deposits. They call them 'stone babies'. With luck, we find ourselves thinking, if Lisika's twelfth and tiniest piglet isn't reabsorbed – is there still a chance of that? – then this is what will happen.

But even that thought is only with us a day or two, rendered redundant (or almost: there's still the placenta) by a call on Saturday, the sixth day, to tell us that the twelfth piglet has been born at last, already in some state of putrefaction in its (his? her?) long-deflated sac, and not without renewed frustration, it having been a volunteers' day at the sanctuary and one of those volunteers having turned up with a dog just as, in the back of the van, Dora, arrived only moments before with Lisika from the vet, was discovering this late birth in process. The dog barked in excitement and the foetus was drawn back. It took a further two hours before it re-presented and birth finally occurred. No stone piglet then, but the placenta still undelivered. It's now a waiting game. Lisika still has a long way ahead of her. The vet is now talking about a hysterectomy. The world, as I say, is a dark poem. We'll see.

Nothing, for almost a month. Alin and Dora, we can only imagine, are caught up with other crises – there is no shortage of them when living with other animals – and we, at the other end of the country, have more than enough of our own work to attend to. But then, with only a week before our return to Australia, we call, to ask how she is. 'Come and see!' is the response, and so, two days later, we do; stop, first, at the micro-sanctuary, to discuss arrangements for a new pig, brought in having fallen from the back of a truck and injured her spine, and then, in the afternoon, drive up the mountain to Hope.

Dora greets us and takes us, past the barn and the house, to where Alin is working on a new shelter in the pig enclosure. Lisika is there with a couple of her offspring: the other two are off in the paddock somewhere. At first it's hard to distinguish her, but that's in part because she is half-immersed in the mud-bath, deeply engaged with an apple. We call and coax – the others come to us – but for her own obscure reasons she won't come near. She's been getting better every day, Alin tells us. You might almost think she hasn't been ill at all. Her appetite's back. They've never seen her so well, he says. We ask about the placenta, but as far as they know it's never appeared. Reabsorbed, they think. Or lost somewhere in the mud.

I imagine it turning slowly to stone.

# The Wound

ALTHOUGH IT HAS TRADITIONALLY BEEN SEEN rather as a product of a distancing from the natural world attendant upon the birth of civilisation (or, in some other speculations, a shift from the matriarchal to the patriarchal construction of the latter), the concept of a wound or rupture deep in the human psyche appears to be as ancient as consciousness itself and the salving of that wound or rupture fundamental to all major religions. With the challenges to traditional religious faith and other master narratives in and after the mid nineteenth century we have been left, increasingly, with the need to reorientate ourselves towards this wound.

One of the first major endeavours in this regard has been in psychology, more specifically in Freud's attempts, in *Civilisation and Its Discontents* (1930) and elsewhere, to articulate more closely the wounding sustained – the constriction of a (putative) wider individual psyche, chiefly through the curtailment of socially unacceptable drives – in the process of enculturation: the concept of an Oedipal crisis, a narrowing into language, the pleasure principle maturing in the reality principle, etc.

Although the concept of a *wound* in the psyche – a sink-hole in the centre of the mind's attempts to explain itself – has not been a comfortable one for their discipline, certain philosophers, too, have tried in some measure to admit and explain it. Nietzsche refers obliquely to this wound when he claims that 'Almost all the problems of philosophy ... pose the same form of question as they did two thousand years ago: how can something originate in its opposite, for example rationality in irrationality, the sentient in the dead, logic in unlogic', etc.[1] For Heidegger there is a *wound in being* caused by thought itself.[2] For Blanchot,[3] as perhaps also for Heidegger, a wound is opened in philosophy by the too-close encounter with political processes and untractable historical realities. Others (Kristeva, Lacan) recast Freud within Structuralism's linguistic metaphor, to see the wound as a matter and product of language – the *Non du père*,[4] language's 'hollowing of being into desire'.[5] For Georges Bataille the wound, as focused in (and in a sense created *by*) the act of communication, is that lack which at once defines one *as* human (*L'homme est ce qui lui manqué*[6]) and allows one to expand into the territory of the lacked.

For Hélène Cixous, who takes this in a more personal direction, the wound is any of a set of traumatic life experiences which, inexplicable, or perhaps just unexplained, call forth writing as a *scarring* process (as in stitching, healing over, masking).[7] And for Jacques Derrida,

---

1    *Human, All Too Human* (1996 [1878], §1), trans. R.J. Hollingdale.
2    An immensely reductive statement, but see, for example, Appelbaum (2009), 'Presence (*parousia*) is necessarily a rupture, an awaking from a latency that is night. Thought is wounded in being thought, the transgression of the logos, which is possibly the kerygmatic idea of original sin. The subject bears consciousness as a wound inherited from birth' (52) – or, more simply, the statement Blanchot gives to his eponymous anti-hero in *Thomas the Obscure*: 'I think, therefore I am not'.
3    'Nazism and Heidegger, this is a wound in thought itself, and each of us is profoundly wounded': 'Our Clandestine Companion' (1980), 43.
4    Implicit in Lacan's seminar on The Psychoses (1955–56), later articulated by Foucault (in *Le 'non' du père*, 1962).
5    Eagleton (1983), 167–8, summarising Lacan, a formulation that appears never to have been made quite so succinctly by Lacan himself.
6    'Man is what he's missed', *Oeuvres completes*, vol. II (Gallimard, 1970), 419.
7    'Preface', *Stigmata*: 'The texts collected and stitched together sewn and resewn in this volume share the trace of a wound. They were caused by a blow, they are the transfiguration of a spilling of blood, be it real or translated into a haemorrhage of the soul ... [Each] is a scene of flight in the face of the intolerable. But not only flight in order to "save one's skin" as French idiom says. In fleeing, the flight saves the trace of what it flees.'

characteristically protean, the wound is, somewhat arbitrarily (although in a manner that might take us back to Nietzsche), one such traumatic experience which one has chosen or has been led to accept as foundational to one's own psyche. In 'Circumfession', for example, he states that all his work has centred about and been drawn from the wound of his own circumcision,[8] although one might assume that, for someone who has based so much of his thought about the concept of phallogocentricity, 'circumcision' here has a significant metaphoric dimension.

The range of such approaches, interesting and stimulating as it might be, has something of the quality of a set of mis-translations of an original just out of reach. Derrida's in particular seems coyly conscious – but how to tell? – that the wound he has chosen is in fact an 'excuse' wound[9] that covers another that for some unstated reason (though I think a deep reading of 'Circumfession' might uncover it) he would rather not approach or expose. The aporetic nature of Cixous' *scarring* (healing on the one hand, but on the other masking a further sign to be read) would seem to suggest something similar.

'Circumfession' is to be found in a book entitled *Jacques Derrida*, each page of which is divided between, on top, a continuous text by Geoffrey Bennington outlining the key concepts and stages of Derrida's thought – a critical biography of sorts – and below, as if a set of footnotes to the above, but only *as if* since this is itself also a continuous and perhaps (rhetorically) counter-text, an account by Derrida himself of what we might call the depth-psychological basis of those concepts and that thought: the psychic drama which has provoked and informed them – or, at least, the psychic drama that, at this point in his life, it is seeming to Derrida has informed them, or (Derrida is a rhetorician, after all) that Derrida is wishing to seem to be thinking has informed them.

It also appears to be the case that Derrida, at the point in his life in which he positions the writing of 'Circumfession', is both suffering

---

8    Geoffrey Bennington/Jacques Derrida, *Jacques Derrida* (1993), 70. Derrida, in *The Animal That Therefore I Am*, says much the same thing of his writing about animals.

9    A term not used by Derrida. I owe it to Michelle Hamadache, of Macquarie University, in correspondence.

from Bell's palsy or something very similar (Lyme disease?), and, simultaneously, undergoing the experience of his mother's dying. Perhaps a touch hypochondriacally, the palsy seems to him to be a beginning to or sign of his own dying. He has one eye which he cannot close – a metaphor? to indicate what, that there is a part of his subject upon which he would normally close his eye but now cannot? (he writes of having one side of his face in the territory of death, the other in life) – and when he drinks water it trickles out of the corner of his mouth, just as it does from the mouth of his mother.[10] He appears, in fact, to be experiencing a deep anxiety that he might (or might not!) die before she does.[11]

His dying mother and the foreshadowing of his own death form a strong current here, albeit only one of several. Another concerns circumcision, a sort of terror and appal at it, and a struggle to make meaning of or with it, though whether this struggle is initiated by Derrida's own circumcision, which it's unlikely he can consciously remember (it would have taken place when he was seven or eight days old), or by a memory of the circumcision of a brother born after him is not clear. There is much in this text about *cuts* and *scars*. There is much about a long, bloodied bandage supposedly unwound from his brother's penis.[12] There is much, too, about a *hidden name*, that further, 'secret' name given to a Jewish boy at the time of his circumcision. Derrida's hidden name – but now he reveals it! – was 'Elie', from 'Elijah', the Hebrew prophet invoked during the circumcision ritual, although ironically this name has never subsequently appeared on Derrida's papers and has therefore (until this) remained hidden.[13]

---

10    Things detailed at 19:98, that is, section 19 of 'Circumfession', on page 98 of the Bennington/Derrida.
11    As it happens – and within the 'duration' of the text itself – the palsy passes, and his mother experiences a (temporary) recovery.
12    Much in this to explain the terror of 'that horrid hole' and the castration anxiety of the encounter with the cat in *The Animal That Therefore I Am*. See 'Meeting Place' above.
13    Much therefore in this, too, then, as in the figure of the absent father here, to support my account of the Derrida whose work serves to protect the unnameable. See 'For the Animals' (2017): 'There's a … deep Abrahamic secret to Derrida, observed by some at the nascence of deconstruction, but

More to the point there is an argument here about what appears to be a sense of guilt established in the infant by the cruelty of the circumcision, as if to say *If I am punished thus, then I must have transgressed in some way*, and a claim or admission that Derrida has constructed from such factors as the above, and perhaps from others that he does not name, a kind of primal wound that, on the one hand, 'explains' the scars he ( Cixous-like) finds himself – and his work – to be marked with, and on the other serves, symbolically and otherwise, and if a little obscurely, as the (only possible) beginning of meaning itself: a wound that, through taking the blame for something for which one may in fact be blameless, one at one and the same time consolidates *and perhaps obscures* by inflicting a further, consequential wound upon oneself.

Intriguing as such a position is, there is an ingenuousness and implausibility about it that one cannot, with any comfort, pass over. On the one hand, even if, and this seems unlikely, so young an infant would remember the injury at all, Derrida's explaining it thus presupposes a sense of transgression and punishment even less likely to have developed at such an age. On the other, such an explanation entails the rather unquestioning acceptance, as *truth*, of what is in fact no more than a *theory* of textual origin (text as scar), and the consequent and very debatable assumption that where there is text there must be wound.

Brushing aside such reservations (but does he?), Derrida presents this self-generating mechanism, in 'Circumfession', nowhere more clearly than in his presentation, as the work nears its end, of a dream in which, 'during a private conversation in an underground place', he was explaining to Jean-Pierre Vernant

---

obscured in the welter of his writing since. A reluctance or refusal to utter the name of Yahweh. Or perhaps better to say to shield it. That behind what these early commentators liked to characterise as deconstruction's *destruction* of meaning itself, its refusal to prioritise, in a text, one meaning over another, was actually an attempt to *protect* some further, deeper meaning – a meaning that, as perhaps his processes of deconstruction attempt to demonstrate, we haven't the capacity to comprehend.' (164–5)

that if he was accusing himself of not being without responsibility for the death of the child, because at its birth he had nipped its neck very gently, this was a way of giving some sense to something that had none, then, half-awake, I understood more clearly the extent to which avowal, even for a crime not committed, simply secretes meaning and order, an intelligibility that arrests, by finally agreeing to confine, the asubjective and endless culpability of chaos. (56:295–6)

We might see this in various ways, one of them as a kind of Ur myth of the creation, out of nowhere, of a drive for meaning, a wound which, assuming its origin, provides the means of capping 'the asubjective and endless culpability of chaos' with an arbitrary avowal – a taking of responsibility – that can then serve as a platform upon which to commence the construction of order. Sense out of senselessness; light out of nescience; morality out of meaningless flux: this might return us yet again to the Nietzsche already quoted – and perhaps, most significantly, it does – although, should we shift our perspective only slightly, the more pertinent voice here might be of Leonardo da Vinci, albeit through Sigmund Freud.

Jean-Pierre Vernant has dreamt, it would seem, that he has killed a newborn child by *nipping its neck*. One thinks, immediately, of the circumcision upon which Derrida's essay turns – the neck as the phallus, the nip as the wound inflicted thereupon – but it's hard not to let the mind then move to Freud's *Leonardo da Vinci, A Memory of His Childhood* (1910), in which he recounts Leonardo's memory, from very early in his infancy, of a visitation, to his crib, of a vulture, who brushed the child's lips with the feathers of his tail.[14]

Freud's interpretation is, essentially, homoerotic (the feather/mouth image one of fellatio, etc.), though it does have other aspects (the vulture summons Mut, the Egyptian vulture-god, female but portrayed with a penis), but while these, and the readings of Leonardo's works (those concerning the Virgin, Saint Anne and the Child in particular)

---

14 While Freud, perhaps intentionally, does not give the vulture a gender (his concern being to conflate the vulture and the mother's nipple), Leonardo indicates quite clearly the vulture was male.

that Freud bases upon them, do engage with Derrida's own preoccupation with his mother, I think they are accessory and that the real summoning here is of Freud's comments concerning the status of Leonardo's dream as such:

> We have here an infantile memory ... of the strangest sort. It is strange on account of its content and account of the time of life in which it was fixed. That a person could retain a memory of the nursing period is perhaps not impossible, but it can in no way be taken as certain. But what this memory of Leonardo states, namely, that a vulture opened the child's mouth with its tail, sounds so improbable, so fabulous, that another conception which puts an end to the two difficulties with one stroke appeals much more to our judgment. The scene of the vulture is not a memory of Leonardo, but a phantasy which he formed later, and transferred into his childhood. The childhood memories of persons often have no different origin, as a matter of fact, they are not fixated from an experience like the conscious memories from the time of maturity and then repeated, but they are not produced until a later period when childhood is already past, they are then changed and disguised and put in the service of later tendencies

– an account, in effect, of the genesis and operation of what I've been calling the excuse wound; a fiction created as a means of 'explaining' and, in that effect, scarring over, a wound for which one has been able to recover no more authentic explanation.

There are, too, some other currents – threads in the braid of this text – that should be mentioned before we go further. We are shown the process of scar-making, that is to say, scars therefore becoming, where they are encountered (Derrida's mother is 'covered' with scars), evidence of further coverings, further occlusions. We are told that an older brother died in infancy a year or two before Derrida was born, and that another brother (he of the bandages) died, at the age of two, when Derrida was ten, so that somewhere in this psychic puzzle there might also be the (guiltless) guilt of the survivor.[15]

We are told too – something which Derrida presents obliquely and fragmentarily, from various sources, as if to bolster a feeling of his own – that it is often or usually the mother who initiates her child's circumcision, indeed in some periods and cultures has actually performed it herself, even eating the foreskin, or drinking it with a mixture of wine and blood, details presented in such a manner as might suggest that, although circumcision is not castration per se, there are some ways in which, for Derrida, it might as well be.

So to, and for, one's open wounds there can be fashioned this kind of explanation. They are in one's past. They are things that have happened to one. And throughout his 'confession' of them (Derrida circles about the idea that in the 'Circumfession' he is at once confessing to and attempting to take – impossible, since she is unable to speak – the confession of his mother), Derrida has summoned, and sometimes confused them (and us) with, a set of further, related details from the confessions of St Augustine of Hippo[16] and others, so that along with the personal there accumulate historical and cultural dimensions – my suspicion confirmed thus that these vectors (the personal, and the historico-cultural) form in fact the arms of an aporia, a contrived explanation to cover (or that happens to cover) something else. They are, to reiterate, an *excuse wound*, serving to cauterise the attempt to plumb a deeper rupture, a default turning into the self that, ironically, closes off further exploration. In these turns – the factors in his current situation that I have outlined – there is, moreover, an unsettlingly familiar, we might even say Freudian, narrative course, which takes emotions and experiences that might encourage a more open searching and cuts them off, folds them back into the encultured/anthropological self, takes away their capacity for development, change, or deeper healing.[17]

---

15   A guilt which, in the microcosm of the one life, reflects – or may be intended by Derrida to reflect – the broader Jewish guilt for survival of the Shoah.

16   Contemporary Annaba (Algeria).

17   There is even, I was a little alarmed to find, that essentially blind (i.e. unexplained and leading nowhere) reference to an Australian Indigenous tribe that has become a kind of trope in modern(ist) thought (Pound, Freud, etc.). In this case, though the specificity is perhaps less significant than the trope itself, the reference (59: 313) is to the Wonghi (also known as the Wangaaypuwan, a clan of the Ngiyampaa nation), doubtless via Frazer's description of Aboriginal circumcision rituals in *The Golden Bough*: 'In the

I am writing this up (or out) like this – this account of Derrida's 'Circumfession' – in some large part because, perhaps all too evidently, that text has moved and disturbed me. It has moved and disturbed me, I think, because at more than a few points in that text I felt that I might almost have been reading about myself. Whether this is a common feeling amongst men who read 'Circumfession' I cannot say. I have only my own reactions to go by.

My own mother died when I was eighteen[18] and although, after only a few years, I might have thought that I was over the grief caused by her death, I think in truth that that grief, acknowledged and unacknowledged, conscious and unconscious, has been a factor in much of my life since. It is also the case that, while not Jewish, I too am circumcised, and although I have absolutely no memory of that occurrence, I have sometimes – not often – wondered if that early wounding might have left some deep scar, some psychic trace nonetheless. And while I am not aware of having had any brother or sister born and deceased before me, indeed am fairly sure there was none (although my father *had* been married before), I am all too aware – was made so, by my mother herself – that I was 'the reason' that there were none born after me: simultaneously with my birth by caesarean section, my mother had a complete hysterectomy, entirely at the doctor's discretion and quite without her prior knowledge or permission.[19]

But this is not a *cir*-Circumfession. My point is that, *un*like Derrida, or so it appears, I have found myself increasingly wary of

---

Wonghi or Wonghibon tribe of New South Wales the youths on approaching manhood are initiated at a secret ceremony, which none but initiated men may witness. Part of the proceedings consists in … giving a new name to the novice, indicative of the change from youth to manhood … It is given out that the youths are each met in turn by a mythical being called Thuremlin (more commonly known as Daramulun), who takes the youth to a distance, kills him, and in some instances cuts him up, after which he restores him to life and knocks out a tooth' (906).

18　And my own father fifteen years later, a reversal of Derrida's order (his father died in 1970, well before his mother).

19　Pointing to the deep scar on her abdomen, asking what it was: 'You did that to me'.

resorting to such matters in my distant past as a means of explaining the various 'wounds' and darknesses of (my) life. It is not just that these kinds of explanation have come to seem trite and their narratives to be dusty and over-worn paths, or that the 'science' of them, when closely interrogated, appears so tenuous. It is also their Narcissism – their Narcissism and a convenient closure that they provide that has come to seem more like a wilful cultural occlusion, a scar-tissue to cover a deeper wound and prevent its further exploration. If there is an upheaval in our psychic lives, as clearly there is in Derrida's at the time of his composing his 'Circumfessions', such matters – such remote and nebulous traumas – provide a convenient place to park it.

Ultimately, I think it can be argued, the manner in which they at once centre us around a lack (wound) and *ground* our identity and individuation there, forms one more part of the sympathetic psycho-technology of saturation capitalism, but that argument is for a different place.[20] It must suffice to say here that the default they provide helps to seal us within ourselves individually, culturally and anthropologically. When we receive a call from outside – or perhaps the deep inside that is, ultimately, a face of the outside – they help to make sure that we do not hear it.

My own position is that our deeper wound – the wound that this inward and Narcissistic process, this *excuse* wound, obscures and so helps us to avoid confronting – has been occasioned by the false and destructive dominion we have given ourselves over the 'natural' world, and over our fellow animals in particular, a deep wound kept as open as it is obscured by our continuing betrayal of and cruelty towards those fellow animals and the manner in particular in which we slaughter and devour what are, in so many ways, our own kind, a wound which, despite ourselves, we keep daily open and must daily occlude.

For many, of course, this will seem to go too far. To those – as sad inducement (not concession) – I would offer instead this thought, an idea on the way to another, that those emotions, those strange and very powerful *desires* (for what are these confusions but deep desires entangled?), that have led us to contrive these excuse wounds to explain

---

20   See my 'The Fallacies' above, where some of this argument is made.

and contain them, are a force the true shape and potential of which we will not know until we cease to neutralise it in this way.[21]

The actuality, contrivance or assumption of a wound *gives us an excuse* to conduct our lives in a certain way. It does not give an excuse to behave as we know or feel we *should*, since this should need no excuse; rather it allows us to behave in a way that in some measure *deviates* from such a path. We know, for example (this, while perhaps the most important example, is still only example) that we should not slaughter or cause suffering in order to supply our own appetite – the child in us cries out at this, when we are forced to face it – but we stubbornly maintain our deviation, find a woundedness upon which we can then base a whole machinery of excuse (wounded, outrageously, undeservedly, we can then take revenge[22]).

At the same time, and hardly coincidentally (it is part and parcel), the wound underpins our Narcissism, gives us further excuse to focus upon, prioritise, validate and valorise the self, and to turn our attention from others. To some extent, of course, we choose and/or construct the wound to suit us and our further purposes (as Freud, psychoanalysis, and the 'cure' of story: the cure lies in finding or contriving a story that suits us, fits the shape of our trauma as we understand it – and as likely as not shapes or re-shapes the trauma as it fits it).

'CIRCUMFESSION' IS AS DENSELY POETIC as it is visceral. Any inclination we might have to leave this strange, dark poem uninterpreted would have to be reconciled with Derrida's very clear assertion that all his work, all his writing, his *project*, has turned about, been underpinned by, or been in explanation and elaboration of one of its key images –

---

21  '*Ist sie den Liebenden leichter? / Ach, sie verdecken sic nur mit einander ihr Los.*' ('Is it any less difficult for lovers? / But they keep on using each other to hide their own fate.') Rilke, 'Second elegy', tr. Mitchell.

22  We may even wish to take revenge upon that which has 'caused' us to wound ourselves. We slaughter animals for our food; we hate/are ashamed of the fact that we have slaughtered them for our food; we take revenge upon them for having caused us shame: a kind of referred guilt that, for example, we see all too commonly at the site of road accidents, in cases where it is the driver who has *caused* the accident who becomes enraged at the innocent party.

> Circumcision – I have never spoken of anything but that, consider the discourse on the limit, the margins, marks, marches, etc., the closure, the ring (alliance and gift), the sacrifice, the bodily inscription, the *pharmakos* excluded or cut off, the cutting/sewing of *Glas* (70)[23]

– an assertion which, as much as its personal dimension, has its symbolic, cultural and intellectually constructed connotations. If the Logos – that which he attempts to explore and understand – is phallocentric, if the phallus is its most stable and enduring symbol, then it is a *wounded* phallus, an *incomplete* phallus, a phallus from which something has been cut (as, but perhaps not quite to the extent of, a ring-barked tree [our *arborescent* epistemology]), a handicapped, limited device, unable to achieve what it strives for.

Another key image, adapted from the end of the seventh book of St Augustine's *Confessions*, in which it configures the physical, created world pervaded by divine spirit,[24] is of a vast sea in which floats an enormous sponge. It appears to be Derrida's addition that this sea is a sea of blood, and that it is therefore blood with which the huge, co-extensive sponge is saturated.

---

23   Although almost no non-human animal is mentioned in 'Circumfession', it's interesting that Derrida makes a similar claim in *The Animal That Therefore I Am*, albeit with some equivocation in his phrasing. 'I will not go back over arguments', he writes there, 'that for a very long time, since I began writing, in fact, I have sought to dedicate to the question of the living and of the living animal. For me that will always have been the most important and decisive question. I have addressed it a thousand times, either directly or obliquely, by means of readings of *all* the philosophers I have taken an interest in' (402). *Glas* (1974), for those unfamiliar with it, counterposes, in separate columns (cum 'Circumfession'), with marginalia, a reading of Hegel with a reading of Genet's autobiographical 'What was left of a Rembrandt cut into small regular squares and tossed into the toilet'.

24   "'It was like a sea, everywhere and in all directions spreading through immense space, simply an infinite sea. And it had in it a great sponge, which was finite, however, and this sponge was filled ... in every part with the immense sea. In this way, I conceived of Thy finite creation as finite [yet full of Thy infinity]" (VII, v, 7)', 'Circumfession' 20:105–6.

This image, of the world as a sponge full of blood; a world of which every part is imbued and heavy with blood; which I take to be an object correlative of something like grief- or wound-consciousness, is a very powerful one,[25] a poetic image (we might even speak of it as symbolising the bruise of poetry itself) succeeding where a discursive explanation might struggle. But what is it doing here? How does it inflect the rest?[26] Elsewhere in 'Circumfession' Derrida writes of using the bandages that have been around the circumcised penis to 'sponge' the blood, and of 'the wound of circumcision' ('in which I return to myself, gather myself, cultivate and colonise hell') as both *escarre* and sponge ('G., listen, it sponges endlessly the blood it expresses' [103–4]), as if the wound of circumcision – circumcision *as* the Wound – were so pervasive and (dare one say?) omnivorous that it coloured the world itself, that nothing can be seen but through that lens.

And yet, of course, it *is* a scar – a wound in itself, but also something that *seals* a wound. 'Circumfession' is riddled with images of skin, of ruptures of or lesions therein, and (at the same time) of the *world* itself as a skin.[27] The mind, too, has a skin. Ruptures in the world's skin – and the mind's – are images, object correlatives, of ruptures in our understanding; are, arguably, invitations.[28] Derrida's mother, therefore, with the bedsores, ruptures in her skin, most readily on her *sacrum*,[29] is also metaphor. As if – a tentative here – there were, in this wound story, a story behind the story Derrida contrives for us:

---

25  Not least because, from my own part-Mediterranean childhood, I remember, vividly, the huge sea-sponges at bath-time.

26  How can we press from our minds the thought of amniotic fluid, the placenta?

27  See, for example, 16:82: 'the *escarre* [bedsore], an archipelago of red and blackish volcanoes, enflamed wounds, crusts and craters, signifiers like wells several centimeters deep, opening here, closing there, on her heels, her hips and sacrum, the very flesh exhibited in its inside, no more secret, no more skin'.

28  See my 'Scheherazade' (1999), 228–38.

29  The *Os sacrum* or holy bone. The *Oxford English Dictionary* offers us, from A. Monro Anat. Human Bones (ed. 3 [1741]) 192: '*Os sacrum* is so called from being offered as a dainty Bit in Sacrifice'. A further strong current in 'Circumfession' is sacrifice, the *making sacred* (think *artifice*), which is to say relegation to the territory of the untouchable, the unquestionable, the placing

a wound behind or beyond the wound he seems to be telling us of. Is he conscious of this, or is this excess, something out of his control? Is he saying/confessing – and is he conscious that he is doing so – that the world is saturated with blood (some two hundred million 'land' animals are killed each day for our appetite[30]), and that in some large part it is this we have contrived our culture to occlude, deny, displace, and distract ourselves from?

The mother – who (in the father's absence? where *is* the father?) initiates the circumcision, and either does it herself or, taking the foreskin (a ring!) and, mixing it with wine and blood that (since Derrida offers and leaves open the possibility that she has performed the *mohel*'s function) she may have sucked from the wound/penis herself, drinks/chews it, in a kind of ritualistic, Oedipal marriage, further shutting out the father (/reason) – is covered with scars; Derrida's text is replete with references to bedsores, scar-covered, some of them re-opening, exposing wounds that are centimetres deep.

Real mother; metaphoric mother: is Derrida imaging, in this way, wounds – scars – that occlude a greater wound or wounds, in the world's body? We can't with any confidence say that he isn't, although by the logic of – under the regime of – the sponge, any blood which seeps from these wounds would be circumcision blood, and the hermeneutic circle – that scar – would close again, like a wagon-train against attack, drawing battle-lines.

Put this all together and one gets what? The mother, as world/nature/physical being, as at once wounding and marrying the child, in such a manner as handicaps (with his *animality*) his ability to know and therefore reach the father?

In a book on *The Jewish Derrida*[31] I find the following, paraphrasing *Glas* which in turn paraphrases Hegel:

---

beyond the bounds of scrutiny and analysis. See my discussion of the sacred in 'The Smoking Vegetarian', above.

30 One of the more conservative estimates available. The figure increases dramatically when we include fish.

31 Ofrat (2001).

prior to the Flood, man lived in harmony with nature, his mother. This idyll was disrupted by the Flood. Henceforth, Mother Nature would turn against mankind. The sole way of overcoming the fury of the rising waters was by taming nature through the power of thought ...

The Jew's attitude towards nature ... is harsh and unfriendly. ... The Jew remains divided and torn ... [A]s illustrated likewise by the annals of the patriarch Abraham, his departure from Ur of the Chaldees in effect cut him off from family, from natural love. The act of separation is thus the very essence of the birth of the Jewish people. Abraham's story is a tale of divorce from nature, of eternal alienness: desert, wilderness, wandering, war with the peoples whose domains he invaded. (48)

Just so. Although it is hardly a Jewish matter, nor even a matter of Abrahamic religions, alone, and I present the above only as a key and symptom. By all that I have been paraphrasing, Derrida means to hold the blood-soaked sponge to represent the manner in which *all* thought, which is to say *thought itself*, is impregnated with this sense of a wounding-by-nature, and that the mother – nature herself – handicaps our being; that she has required, in order for us to become what we have become, a separation, a severance from her; that she has required that wounding.

But that is only the excuse wound, the wound that hides the greater wound from us, a greater wound that, paradoxically, is in one sense only a small step away, because it is so daily about us – the earth, soaked with the blood of the slaughtered *kine* (1.4 billion land animals each week ...), let on account of the deep wound in our logos – yet in another sense is so far away, so *buried*, that it is almost inconceivable, a supposition, a figment, perhaps, of this author's imagination. Certainly (but it is fifteen years before the *turn* of *The Animal That Therefore I Am*) no non-human animal is mentioned at any point in 'Circumfession'. Whether – from this master of the *trace* – that need mean that such animals are not there is another matter: in one sense 'Circumfession' is an account of its author's powerful and dis-placing encounter with his own animality, even if (in this text about hidden names) he never expresses it as such. Later, in *The Animal That Therefore I Am*, Derrida will assert that, all his (critical) life, he has been writing about animals. In 'Circumfession' he claims that, through all that he has written so far, *circumcision* has been his central, guiding obsession. In all likelihood our inclination will be to take the similarity between these two statements as no more than a rhetorical tic. But what if we were to superimpose them? What if there were a step further we might take, to a point from which we might see that they are in effect the same statement?

And what then – now – to do? *Take* that step, cut through the scars, attempt to reach, through one kind of wound, to the other? Or, acknowledging that the wound is everywhere, simply turn, have done with that inward vector, walk away, all the freer then to commence the immense work of triage, redress and undoing that our ancient war against nature and non-human animals has left us?

# The Rats of Lord Howe

No EVIDENCE HAS YET SURFACED that humans had ever set foot upon the tiny, isolated portion of the inhabitable earth now known as Lord Howe Island (320 nautical miles east of the Australian mainland, at 31°33'15"S) until Lieutenant Henry Lidgbird Ball, commander of H.M.S. *Supply*, led a landing party in early March 1788 to claim it and the surrounding islets as the latest outpost of the British Empire. These rocky outcrops might have lain undisturbed a few years longer had Captain James Cook, on the *Resolution*, not 'discovered' their neighbour (484 nautical miles away) in 1774 and named it after the Duchess of Norfolk.

Cook had noted the distinctive, ship-mast tall pines on Norfolk Island, and the abundance of a plant very like the flax used to make the British navy's sails. By the mid-1780s the Admiralty, its North American supply of masts and European sources of flax effectually cut off, was concerned that Norfolk not fall into French hands. It had, accordingly, instructed Captain Arthur Phillip, leader of the First Fleet, to dispatch a vessel, as soon as he was able, to establish an outpost on Norfolk, doubtless – such was its remoteness – to hold the worst of his felons.

On 14 February, scarcely two weeks after arriving in Sydney Cove, the *Supply*, under Ball's command, left for Norfolk Island, carrying a party of twenty-two under the leadership of Philip Gidley King (who'd later become third Governor of New South Wales) to establish a British outpost and penal settlement as instructed. On 17 February the *Supply* passed an unidentified island which Lieutenant Ball determined to visit on his return journey. This he did on 13 March, naming the place Lord Howe's Island after Richard Howe, First Lord of the Admiralty. Not backward in coming forward, he then named Lord Howe's most prominent summit Mount Lidgbird, and, noting a dramatic spear-head of stone towering five hundred metres out of the ocean a few miles to the south-east, named this Ball's Pyramid.

Perhaps because Ball had found no source of fresh water upon it, Lord Howe remained uninhabited until 1834, when a small party of eight – three sailors with their Maori wives and two Maori boys – set up there. In other respects, however, Ball's account was encouraging enough to occasion a further visit, barely two months later, by four ships from the First Fleet, the *Supply*, the *Scarborough*, the *Lady Penrhyn* and the *Charlotte*, on their way back to England, a visit from which date the first records of the island's extraordinary – and extraordinarily prolific – wildlife: an abundance of turtles, vast flocks of nesting seabirds, land birds so curious and unused to predation that they showed no fear. 'When I was in the woods amongst the Birds', wrote Arthur Bowes, surgeon of the *Lady Penrhyn*, 'I cd not help picturing to myself the Golden Age as described by Ovid.'[1]

But abundance is relative. Teeming with wildlife as it doubtless appeared to be to these first visitors, the island itself is very small (only 14.55 square kilometres, or 3600 acres: the size of a modest sheep station), and the variety of its non-human inhabitants limited. There may, for example, have been fifteen indigenous bird species, but their populations were quintessentially fragile. Doubtless, from a conventional, human perspective, the *Resolution*, the *Supply*, the *Scarborough*, the *Lady Penrhyn* and the *Charlotte* were vessels of discovery, but from the perspective of the wildlife they were death ships.

From Ball's first glimpse of the island as he passed, the doom of the non-humans of Lord Howe was sealed. It might have been forty-six years before it received its first permanent human residents, but Lord Howe, through the intervening decades, was more and more frequently visited,

---

1    'A Journal of a Voyage from Portsmouth to New South Wales and China in the *Lady Penrhyn*', MS, Mitchell Library, quoted in Hindwood (1938), 321.

at first by ships on their way to and from Norfolk, then very quickly, as the industry established itself and the island became a favoured provisioning point, by whaling ships, some fifty or sixty a year by mid-century. Whalers dropped off pigs and goats there, to breed and become a future food source.[2] And of course there were always turtles, birds.

Amongst the most abundant and most popular of the latter, at least initially, were the white gallinule, or Lord Howe Island swamp-hen (*Porphyrio albus*) and *Janthoenus godmanae*, the white-throated Lord Howe pigeon. The gallinule, a large, fulsome bird, was of the *rail* family, and probably flightless. Very curious, and remarkably *tame* (if that's the right word for having no experience of predation), it could be run down or knocked over with a stick. And the pigeon – plumper than any pigeons these early crews had yet seen – could be lifted off the boughs of trees.

By the time the island's first residents stepped ashore, the gallinule was already long extinct (the last recorded sighting is from 1790). By the mid-1850s the pigeon, too, was gone. A third, the Lord Howe parakeet (*Cyanoramphus novaezelandiae subflavescens*), had vanished by 1870 (the last record a pair seen flying at dusk into the forest in 1869), in large part for the damage it had done to settlers' crops, though the cats introduced by whalers in the 1850s can't have helped.

As a consequence of over-fishing, the discovery of kerosene, and the American civil war, the whaling industry, in the last decades of the nineteenth century, declined as rapidly as it had appeared. A penal colony proposed for Lord Howe never eventuated, and that on Norfolk closed in 1855. Mice were introduced to Lord Howe in the 1860s, but their growing presence can't have troubled the human population sufficiently to deter it from dramatically reducing the cat population, and the settlements of both Lord Howe and Norfolk grew. By 1910 there was even a steamship, the *Makambo*, making a regular mail run from Sydney to Lord Howe, Norfolk and the New Hebrides, offering the possibility of tourism for those who had the time and money and were

---

2    A great great great uncle of mine, Richard Copping, was in the crew of a whaler which provisioned there in 1851 and killed enough whales in neighbouring waters to produce 150 barrels of oil.

prepared to take a journey that had clearly not impressed a journalist writing of the island in the *Hobart Mercury* in 1927:

> It is a paradise of graceful palms, of flowering shrubs, giant banyan trees, and coral-studded waters; and each year is becoming increasingly popular for those in search of rest and quietness, provided always that they are not discouraged at the outset by the present inadequate and uncomfortable steamer service between Sydney and the island. Occupying anything from 48 to 58 hours, according to the weather, the voyage is undertaken by the *s.s. Makambo*, of 1,100 tons register which although a good sea boat, has no pretentions to either size or speed, and the restriction of accommodation is often very unpleasant. The steamer calls at the island once in about five weeks.[3]

In service until 1931, the *Makambo* is at once vital to Lord Howe's history during this period, and, owing to an incident in mid-1918, notorious within it.

There was, in 1918, no safe berth at Lord Howe. Ships had then as now to negotiate carefully the reefs around the island before anchoring offshore, either at Ned's Beach on the eastern side or Lagoon Beach on the western, depending upon wind direction and conditions at sea. The night in question – of the 15 June, 1918 – was calm, clear and moonlit, but the *Makambo*'s commander, Captain Weatherall, was so ill with whooping cough that he passed out while negotiating the reef.[4] The ship, with its load of copra and seventeen passengers, ran aground, ripping a hole in her hold. One passenger, a Miss Reardon, returning to Sydney from Norfolk Island, drowned when a surfboat capsized during the evacuation.

The ship itself did not sink, but it took nine days to re-float, during which time a number of rats found their way ashore. With only the few remaining boobook owls to keep their numbers down and an abundance of prey unused to rodents, they multiplied rapidly.

---

3     M.S.R. Sharland, 4 January, p. 8 (via Trove archive, NLA).
4     This detail from an ABC Radio National piece for 'Off Track' by Ann Jones. See https://ab.co/3htYjGW (posted 2 June 2017).

The damage was almost immediate. 'But two years ago [1919]', writes one witness,[5]

> [t]he forest of Lord Howe Island was joyous with the notes of myriads of birds, large and small and of many kinds ... They were unmolested save by each other, the residents of the island rarely disturbing their harmony. To-day, however, the ravages of rats, the worst enemy of mankind, which have been accidentally introduced, have made the note of a bird rare, and the sight of one, save the strong-billed Magpie and the Kingfisher (*halcyon*), even rarer. Within two years this paradise of birds has become a wilderness, and the quiet of death reigns where all was melody.

By the mid-1930s, or very shortly thereafter, five more unique native birds, the Lord Howe fly-eater, the Lord Howe fantail, the robust silver-eye, the Lord Howe starling (or 'Cudgirmaruk'), and the ouzel or 'doctor bird',[6] had gone the way of the gallinule, the pigeon and the parakeet. Granted, some of these – for example the ouzel (also known as the vinous-tinted thrush), which had been common as recently as 1906 – had already significantly declined due to human disturbance and the predation of other introduced species, but the arrival of the rats seems to have been their death knell.

Perhaps the most famous victim of these infamous migrants was *Dryococelus australis*, the Lord Howe stick insect, known to the island's human inhabitants as the 'tree lobster', a phasmid so large (it could grow to a length of fifteen centimetres) and chunky that islanders are reported to have been in the habit of pulling its legs off to bait their fish hooks.

Once plentiful – though the cats and pigs can't have helped their numbers – the rats seem to have rapidly wiped these insects out. For much of the twentieth century they were thought extinct – the last recorded sightings were from the early 1920s – until, in 1964, climbers on Ball's Pyramid found a few seemingly fresh carcases, and an

---

5    Allan R. McCulloch (1921), 43.
6    Its plumage resembling the Quaker-brown coat of Dr John Foulis, who lived on the island from 1844–47.

expedition dedicated to resolving the matter came upon a colony estimated to be of around twenty-four individuals in a small patch of Melaleuca approximately a hundred metres up the rock-face, and a subsequent expedition (2003) managed to bring back a pair ('Adam and Eve') in order to establish a breeding program at Melbourne Zoo.

For a time, discovered at the very brink of extinction, the tree lobster was regarded as the world's rarest insect. Though the breeding program has been very successful – there are now 13,000 or more[7] – they are as yet still a captive population and there are (of course) hopes that, if the rat problem can somehow be solved, they might one day return to Lord Howe Island. Doubtless there are hopes, also – certainly this author hopes – that there is still a colony on Ball's Pyramid, clinging to life on that rocky pinnacle (and to one another, as apparently they do: 'The males and females', I read, 'form some kind of a bond … During the night the couple sleeps together with three of the male's legs wrapped around the female'[8]).

Lord Howe islanders have been dealing with rats now for just on a hundred years – with manual traps, with fox terriers, with the various generations of poisons that the century has provided. For a time a bounty system was in place – a few pennies per tail, paid by the Island Board. In 1928, K.A. Hindwood records, bounty was paid on 21,214 tails. Even raptors have been enlisted. 'In an effort to check the increase of the rats', writes Hindwood, 'almost one hundred owls of several kinds were sent to the island between 1922 and 1930' (322–23). There was already a native owl on the island – *Ninox novaeseelandiae albaria,* the Lord Howe boobook (or morepork, a subspecies of *Ninox novaeseelandiae,* the Southern boobook) – but the arrival of the imported owls appears to have hastened this bird's disappearance. The precise point of its extinction is hard to determine. It was still present in 1920, and, since boobooks were heard as late as 1950, it may even have survived another thirty years, but some of the owls introduced in the '20s were Southern boobooks, captured near Sydney, and these calls may have come from them.

---

7    A 2016 figure. See the Wikipedia entry for *Dryococelus australis.*
8    Lewis (2010) attributes this information to Wikipedia, from which it would seem to have been removed.

There are still owls on Lord Howe, or were at the time (June 2019) this essay was begun: twenty to fifty breeding pairs. Although, along with the Southern boobooks, there were also American barn owls and some great horned owls, the bulk of the birds imported in the 1920s were Tasmanian masked owls (*Tyto novaehollandiae castanops*). These owls were the only ones to actually establish themselves, and Lord Howe is now listed as a second precinct of the species – ecologically precious, one would think, given that in its native Tasmania the masked owl is listed variously as vulnerable and endangered, with an estimated population reduced to less than five hundred breeding pairs.

Tasmanian masked owls were first described by John Gould, who wrote, in his *Handbook of Australian Birds* (1865), of their 'great size and powerful form': 'Probably few of the Raptorial birds,' he said, 'with the exception of the Eagles, are more formidable or more sanguinary in disposition' (62–3).

Not sanguine enough, however, to curb, alone, the number of rats on the island. While there's no doubt they've played their part (rodents are their main diet), the owls – not that it was ever their choice to come in the first place – are considered an invasive species on Lord Howe and, rare as they are, have sometimes been culled there.[9]

Just how many rats might now be on Lord Howe is unclear. Recent estimations put 360,000 rodents there (250,000 mice, 110,000 rats). If rats and mice were to be evenly distributed over its surface (and of course they're not: apparently the greatest concentration is on the top of Mt Gower), this would mean one rat for every 132.3 square metres, and one mouse for every 58.2.

Whether this constitutes an invasion I'm not sure. And of course it rests on our acceptance of the estimations in the first place. Much of the island is of scrub so thick, apparently, that one can barely crawl through it. Mice live in burrows under the ground and, particularly in such terrain, are notoriously difficult to find, let alone count. How have these figures been arrived at? One islander speaks of friends who've been coming to the island for sixty years and have never seen a live rat.

---

9   According to Woodford (2010), 108 were killed in a culling program between 1988 and 2006.

As we might imagine, the success of such containment methods and the damage rats do have long been open questions. The five bird extinctions in which their involvement seems clear all occurred between 1919 and 1938,[10] and, of the remaining indigenous birds of the island, the Lord Howe silvereye (as opposed to the *robust* silvereye), the golden whistler, and the green-winged (or 'emerald') dove (in fact no longer considered indigenous, but why exclude it?) exist in large, stable and apparently untroubled populations (5000 silvereye, 2000 golden whistler; I find no figure for the dove).

The Lord Howe wood hen, it's true, was on the verge of extinction in 1980, when it was thought only a dozen remained, but through careful management the numbers have returned to around three hundred, which may in fact be the optimal population for the area. The Lord Howe currawong, whose population, for reasons not clearly understood – rat poison (DDT at that point?)?: currawongs are happy to eat dead rats – had declined to between thirty and fifty in the early 1970s, has also made a strong recovery, to two hundred and fifty or more at the present time.

There have, that is, been no new extinctions since the late 1930s, unless we include the boobook, which is in fact unlikely to have lasted so long, and more likely to have been hunted out by humans and other owls than by rats (which would have been the boobook's prey). The recent story is more of creatures brought *back* from the edge. It may even be possible – the overall rat numbers, Helen Tiffin tells us,[11] appear not to have changed for the last forty years – that a kind of balance or accommodation has been reached.

Recent accounts of Lord Howe, however, portray a place in ecological crisis. Rats have overrun the island, we are told, and are threatening numerous endangered species. Time is running out and something decisive must be done, a final solution found.

How has this crisis come about? And is what is now being done the right, or best, or most effective thing *to* be done?

---

10   Recher (1974), 66.
11   Tiffin (2019).

As I WRITE, HELICOPTERS are delivering, sector by sector, load after load (the figure of forty-two tons is frequently mentioned) of the 'advanced' haemorrhagic rodenticide, brodifacoum, over the entire exposed surface of Lord Howe Island, with the exception of the houses, gardens, and other buildings of the settlement, throughout which, instead, twenty-eight thousand baits of the same rodenticide are being laid manually at distances of ten metres apart. People speak of this as a 'carpet bombing'. I've heard it likened to the use of napalm during the Vietnam War.

The intention of this Rodent Eradication Program (REP) is to eliminate, in their entirety, the mice, rats and owls of Lord Howe. Although, as we've just seen, the rodent numbers seem to have long stabilised, and, for better or for worse, the rats and mice could be argued to have created and now occupy an ecological niche – although the most endangered birds seem to have been brought back from the brink, the last rat-induced extinctions are now eighty to one hundred years in the past, and the rest of the remaining indigenous bird species seem to be in very healthy numbers indeed – three hundred and sixty thousand rodents are to be exterminated.

There will be by-kill. The planners and managers of the REP do admit that. Sufficient numbers of the creatures most at risk, however (the wood hen, the currawong), have been captured in advance and are being kept aside, to be re-introduced when tests show the poison is no longer present. Other natives – the silvereye, the dove, the whistler – may be impacted but it seems to be assumed it's unlikely this will be beyond their capacity to recover. In time, the rats, the mice, the masked owls gone, the fauna and flora of the island will be restored to something like their pre-1788 glory, their centrepiece – the jewel in the island's crown – *Dryococelus australis*, the Lord Howe phasmid, until so recently the rarest insect in the world.

Brodifacoum. 'First generation anti-coagulants,' writes W.F. Benfield, 'have been superseded by the second generation, brodifacoum. ... Similar in their mode of action, the difference is their extreme longevity. An animal may get a small sub-lethal dose and then some time later, another and then another until finally the critical level is reached ... with the victim generally dying from gastric haemorrhage, sometimes a year or more after the first bait was consumed.'[12] 'Brodifacoum', I find on a pest treatment website, in clear contradiction of assurances given by the REP with regard to fish, 'is extremely lethal to rodents, birds, fish, and other small mammals'[13] – whose death is slow and agonising, and whose suffering, for most of us, is beyond imagining.

To those more inclined to look at the natural environment with compassion than as a kind of scientific Lego, this is a debacle, an ecological disaster in the making, an idea bred from the digital age and the concept that, when one's computer begins to slow and clutter, seize up or behave other than one desires, one can back up one's files, press *reset*, return to factory settings and start again. Except, of course, that we are dealing with lives, with species, and with an ecology that in so many ways we are still only just beginning to understand. If anything *is* clear, it should be that ecological resets aren't possible. For starters, and *only* for starters, once a species is gone (test-tube thylacines aside) it's gone; any reboot will have missing components, some of them very likely quite unknown, and will have to limp along without them.

In this regard alone the project is curious. It is advertised primarily as a concern for the preservation, from rodents, of endangered species, principally of *indigenous* species (the first crime of the rats and mice, it would seem, is that they're *non*-indigenous or 'invasive'), and principally (although not entirely: there are also plants, insects, reptiles of concern) of indigenous birds – such, in any case, as are left and appear to have (ironically) *survived* the Invasion of the Rodents.

It is troubling, then, that it's these self-same birds that are most likely to be by-kill of this pesticidal carpeting. Some two hundred of the wood hens and a hundred and twenty-five of the currawongs, nursed back from extinction once already, are being shielded from this reset (the computer file back-up) by the aforementioned program of capture and sequestration being run by staff from Taronga Zoo, but the rest – ground-level seed-eaters on the one hand and, on the other, carnivores who will consume the dying/dead mice and rats – will be collateral damage. As, it seems, will be a certain proportion – how could it be otherwise, since they live on a diet of insects and small

---

12   See Benfield (2015), location 1822.
13   See https://bit.ly/3miaA4P.

seeds? – of the indigenous and now truly endangered silvereyes, and the ground-foraging, seed-eating doves, neither of whom (along, curiously, with the endangered Lord Howe *flax snail*[14]) gets a mention in any protection provisions I've seen.

As for the Tasmanian masked owl, according to a logic I find it difficult to follow (endangered in its home territory, open season if it strays out of it), on Lord Howe an invasive species (although, as I've said, there *were* owls there pre-1788), they're condemned not so much for what they've done as for what they might do – develop a taste for tree lobsters – if ever the rat problem is solved and *Dryococelus australis* is able to return.

Lord Howe is home to far more than its beleaguered natives, and its ecology, like ecologies everywhere, is far more than the sum of its endemic species. What effect will this massive poisoning have on the island's seabirds, or the marine life that surrounds it? The project would seem to assume that if a creature might be reintroduced after the mass poisoning then that creature is temporarily expendable. Seabirds will likely reintroduce themselves. Seabirds, after all, derive their diet *from* the sea, are not *seed*-eaters, and are therefore unlikely to show interest in the bait.

To ensure the death of every rodent on the island, however, the poison must reach quite literally from shore to shore. Brodifacoum mightn't be *placed in* the sea, but the first high tide or shower of rain will get it there. 'One side of the highest point, Mount Gower,' writes Benfield, 'is virtually a cliff plunging from the summit to the sea which will result in a lot of brodifacoum ending up in the water.'[15]

Brodifacoum, we're told, won't affect marine life. It's a haemorrhagic, an anticoagulant: fish systems are different. We won't see fish among its by-kill. But brodifacoum accumulates in the tissue, principally in the

---

14  Benfield (2015), location 1805: 'Until recently, it was thought that the victim must be an animal or bird with a vascular system and, because insects don't have vascular systems, they are supposed to be unaffected. It appears this may not be the case, as an investigation following a recent rat eradication programme at Fregate Island in the Seychelles has shown. Rare snails were severely affected and one species may actually be extinct as a result of brodifacoum poisoning.'

15  Benfield, *The Tasmanian Times* (4 June 2015).

liver. If it's true it won't affect marine life, why is it residents and visitors to the island are being told not to eat island fish – especially the *livers* of fish – until tests show them to be poison free?[16]

Maybe we won't see seabirds amongst the *immediate* by-kill, but brodifacoum accumulating in the tissue of fish will therefore accumulate in the tissue of birds that eat them. Can we be sure that it won't reach a point where it affects the circulatory system of those birds, or where, as have numerous pesticides before it, it affects their health, longevity and reproductivity in other ways? In 1962 Rachel Carson's *Silent Spring* alerted us to the devastating effects of DDT; a key piece of evidence was that pesticide's impact upon the American bald eagle. DDT thinned the shells of their eggs, which then broke in the nests. The pesticide was banned in 1972, and the bald eagle population slowly recovered. Shift to Rat Island, Alaska, 2009, where 'miscalculations' in a brodifacoum project led to the death of over four hundred and thirty of the birds it was supposed to protect, among them forty-six bald eagles.

Exponents of the Lord Howe project tell one story; the history of and stipulated precautions for the poison's application tell another. Consider the provisions concerning poultry and livestock in the permit issued by the Australian Pesticides and Veterinary Medicines Authority. 'DO NOT', the permit holders are told, 'allow livestock and poultry to come into contact with the product'. Livestock must be 'penned in containment areas' with a five-metre buffer zone; 'residents and tourists must be advised not to eat any meat or offal from any animals on LHI during the baiting operation'. Milk from the dairy herd is to be 'disposed of until bait is no longer present and laboratory testing confirms that there are no brodifacoum residues present'. Etcetera.

The project has so many difficulties and unknowns one wonders how it could work at all. The first and greatest is that it is a *human* project in the first place. It promises that, once the poison has done its

---

16  But how, since the action of brodifacoum is cumulative – i.e. builds up in the tissue until a critical point is reached and the animal dies – are fish going to be 'purified' of the poison? Isn't the only way fish at Lord Howe Island are going to be found 'poison free' going to be because all affected fish have died – that, contrary to the project's assurance that they will not be affected, they are *expected* to die?

work, the island will be free of rats forever. But with humans there *is* no forever. The specifications of other brodifacoum saturations speak ominously of the need for 'maintenance' poisonings down the track. And governments change their minds. Legislation changes. Protective measures are neglected or abandoned. A hurricane or far off tsunami might bring to the reef a rat-infested wreck. Or some person might simply make a mistake, or become careless.

More immediately there are (a) the containment of the poison, (b) the paradox that the project must inevitably kill so many of the creatures it's supposed to save, and (c) the almost certainty – so difficult is so much of the terrain – that there'll be parts of the island the poison does not reach. There's also – (e) but where could I find the time and space for this? – *rat*ness itself: the idea that rats cannot learn, cannot suss out what is happening, will be where humans want them to be, will behave as humans expect them to behave, etc. And there is, of course – (f) – a fundamental matter of cooperation.

The project will only work if all human residents fall into line. It would only take one islander to shift, tamper with or remove some baits, harbour a pair of mice or rats, fail to disclose the location of a nest, to jeopardise the entire endeavour. And the islanders have voted for the project by only the slimmest of margins (52 per cent to 48 per cent). It doesn't take much reading between lines to sense that, even to attain this slimmest of majorities, coercion has been involved, and resentment bred. One informed estimate suggests that, at the time the project at last swung into action (June 2019), there remained at least one hundred islanders deeply opposed to it.[17] Ninety-three islanders have signed a protest to the Administrative Appeals Tribunal.

Officially, Lord Howe Island (population 382) is administered by an Island Board, but the ultimate authority in island affairs is the state of New South Wales, which administers the island through its Department of Environment and Heritage. The feeling that islanders might have to some extent been bullied into accepting a project conceived and accepted elsewhere emerges strongly from certain parts of the documentation setting up the project, though perhaps nowhere more so than in the draconian nature of provisions set out for access to

---

17   Tiffin, podcast (2019).

and inspection of islanders' property. 'Where consent is not provided for REP staff to undertake baiting in residential dwellings,' reads a Business Paper for the September 2018 Board meeting,

> [a] Control Order or Biosecurity Direction would direct the owner/occupier to undertake the measures (i.e. baiting) themselves subject to suitable training and verification. ...
>
> However, if there is non-compliance with an Order/ Direction, there are penalties and powers within the Act under both mechanisms for Authorised Officers to enter a residential premise, without consent, to undertake the measures, with DPI authority. The use of these powers is highly sensitive and would only be considered as a last resort where all reasonable avenues to gain consensual access or to have the occupier undertake the measures themselves have been exhausted. The use of this power would only be considered when the entire operation is at high risk of failure if it is not used. It is considered possible that this would need to be used on at least two and up to five properties ...
>
> The Board should note that use of a Control Order/ Biosecurity Direction ... is considered critical for project success.[18]

Whose *bio*? Whose *security*? The New South Wales government's highly contentious *Biosecurity Act 2015*, to which these paragraphs refer, also prevents animal rights activists from visually documenting instances of cruelty towards non-human animals in industrial farming facilities, protecting the *bio* of big business, the *security* of large investments. Ridding a remote island of rats and protecting abattoirs from the prying eye of concerned and compassionate members of the public would seem worlds apart, but are they really? Am I alone in thinking there's a rather dramatic disproportion involved in the killing of 360,000 rodents and an as-yet-incalculable number of members of other animal species, a good many of them members of the self-same endangered species the eradication is supposedly designed to protect, in order, ultimately – for

---

18    https://bit.ly/3mly7BK.

this does seem to be a core issue – to be able to re-introduce one curious insect thought already lost almost a hundred years ago?

*Is* this insect a core issue? Or is *Dryococelus australis* itself a victim here, a pretext, a pawn in a larger game? And what might that game be? The answer is at least double-barrelled. First there is the undoubtable pressure of what Benfield has called *corporate conservation*[19] – a rolling coalition of business, science and government concerned with the marketing, coordination and supply of island rodent exterminations and restoration and the profits therefrom, a coalition for our intents and purposes centred in New Zealand (although Australia is an enthusiastic participant), with its government-owned poisons factory (Animal Control Products), its own research centres (the Invasive Species Specialist Group, the Island Eradication Advisory Group, etc.), and numerous people trained up to design, manage and carry out such projects.

Although the collaborative experience of this coalition is extensive (the figure of 'over 700 island eradications' is frequently cited), it has as yet little experience in conducting such operations on islands with permanent and relatively self-sustaining communities (as opposed to scientific and military outposts, etc.) and the problems of farm and domestic animals, lifelong attachment, investment, exposure and public relations (etc.) that such communities present. Lord Howe, it seems, is a first, a client all the more significant in the light of New Zealand's declared intention (2016) to be 'pest free' by 2050 (in the pursuit of which many, many millions will die: there are over thirty million possums, alone, in New Zealand, let alone cats, rats, deer, foxes, blackbirds and so on), and its interim goal to eradicate rodents from mainland national reserves by 2025. All that's been needed is a 'crisis' to bring the various agencies together, and the population of Lord Howe into line. In this sense the return of *Dryococelus australis* – quite literally

---

19  Benfield (2015). While some of Benfield's inferences trouble me, and while he makes almost no mention of Lord Howe Island in this book, his investigations into the networks enabling rodent eliminations and island restorations I find all too convincing. In almost every aspect Lord Howe's case bears out his thesis.

a *manufactured* crisis – has been a godsend. How can the tree lobster be reintroduced while its natural home is infested by rats?

The second barrel is more conceptual. Corporate conservation is driven – fuelled – by the assumption of the primacy of the native (ironically, a very *speciesist* assumption) over the introduced. And, partly in corporate conservation hands, the concern for the preservation of native species, laudable as in so many respects it might be, has become alarmingly lethal. If a life is a life, how is it we've come to a point where we'll end three hundred and sixty-something thousand lives to preserve thirteen thousand we have laboriously, if very successfully, bred up in a laboratory (and which now have 'insurance' populations in the United States, Canada and England)?

Am I arguing for a choice of rat over tree lobster or vice versa? No. And that's my point. How – by what ideological parameters – have we come to feel we must make such choices in the first place? We eliminate certain species from certain places in the interests of *biodiversity* – are prepared to destroy incalculable intricacies of ecological interaction in order to rid one place or another of some species or another that we humans find repugnant or otherwise undesirable – but what kind of biodiversity is it that requires us to withdraw species *from* it in this way, that requires mass slaughter, *biocide*, and that, in the process, sends ecosystems back, destroys any progress they may have made in this treacherous, anthropocentric terrain, does not allow them to grope their own, albeit very difficult path forward?

We've reached a point, it seems, where the concerns of conservation and the rights of individual non-human beings have become incompatible. But do they *need* to be, or is this, as I've come to suspect, primarily a matter of *economics* (on the one hand our reluctance to spend anything but the bare necessity upon the environment, and on the other the power of corporate conservation and its need for revenue, profit, expansion: what I call 'death by dividend')? Economics, that is, and a fundamental failure, a laziness in our thinking.

Is there no other way out of this than poison?

As for the owls, ah, well. I have not, as I have written this, been able to keep from my mind Kikinda, a town in northern Serbia to the trees in a central square of which, early every winter, hundreds of long-eared owls descend – seven hundred or more in 2006, over five hundred in 2015 – in what is reputed to be the largest winter roosting of such owls in the world. In the five months they are resident there – apart from all the other wonders of their visit – they supposedly eat over half a million rodents, rats and mice both. Too late for Lord Howe, of course, and the long-eared owl may not have been an option, but who knows? All sorts of calculations go through the mind.

# Archons at the End of Time

WE CAN BE FORGIVEN, I THINK, FOR WONDERING, some days, if we're experiencing the end of the world. At least as we've been knowing it. Climate change. The extreme weather events and other alarming phenomena that are attending it. Droughts where there have been none in living memory, floods such as we've not seen before, heat-waves, megafires, tsunamis, earthquakes, the melting of the Arctic and Antarctic icecaps, the disappearance of small island nations in the rising seas, the death of natural wonders. Such things alone might have overwhelmed us, but there is so much more: the relentless and it seems unstoppable human ravaging of natural resources, the destruction of rainforests, extinctions of wildlife, the reaching of peak oil.[1]

We manifest, in the face of such relentless catastrophe, at once an almost inconceivable inertia and a greed so rapacious it might not be inappropriate to term it suicidal. Where leadership and resolution are called for we have political paralysis, corruption and disintegration; instead of cooperation and redress we have regimes of terror. The perverse and destructive systems and assumptions driving our economies, dominating our political landscape, and determining our understanding of worth and achievement (growth, progress,

---

1    And now, as this book goes to press (2021), a pandemic (this essay was written in 2017).

consumption), have so pervaded our psyches we seem powerless to resist them. I could continue – the list is only just beginning (we are consumed by our own consumerism, suffering death by dividend) – but the impact appears only to lessen with each addition to it. Perhaps the most extraordinary development of all is that these things and our attendant numbness are now tantamount to the state of our being, no real news to anybody.

The concept of an end of the world has been with us for a very long time. The Old Norse had their Ragnarök. The Vedas tell of the Yugas (currently the Kali Yuga, or Age of Vice). The Judeo-Christian tradition gives us the Book of Revelations. Armageddon. The Apocalypse with its four horsemen. The Last Judgement. In some versions the process is cyclical: the end of humanity ushers in a new period *of* humanity. What we are now experiencing doesn't appear to be so. Speculations about moving to other planets notwithstanding, it looks as if – saving some almost-inconceivable *volte face* – this end is going to be *the* end. The Doomsday clock ticks. We are only a minute or two from midnight.

Although they've been referring to something rather different, philosophers have for some time now talked of the end of history. It was arguably Hegel who instigated the notion but he had plenty to build upon. Human history, he tells us, is *dialectical*: we move relentlessly, *through* history, towards Truth in a continual process of negation, competing errors and half-truths (a kind of intellectual Darwinism), always on an upward, refining path, as one truth proves stronger than another, and another proves stronger still. The process, as he presents it, will by definition reach an end. If we take this upward, refining process to be history itself (I don't see that we have to), then at that point we will have reached the End of History.

Subsequent philosophy has re-framed this sense of historical ending as a socio-cultural matter, the movement towards, acceptance and attainment of a final form of a particular political, economic or social system. The contemporary debate seems to be over whether or not we've reached that end (a kind of Peak Human) in, say, the capitulation of the last great Communist strongholds to a globalised Capitalism, and for some time now philosophers have been concerning themselves with what comes after.

Of course, it was not, in the mind of Hegel or of most of those who've subsequently concerned themselves with the issue, that the end of history would be as final, catastrophic, and destructive to humanity and nature alike as the ending we now seem to be facing. But we are *homo omnicidens* (killers of everything). We will not, for example, be leaving the world to the other animals: at least, not to all that many. Ours is the great age of extinctions. We seem determined to take as many other species with us – or send them before us – as we can.

It pays to think of them, the other animals, the *non-human* animals, nevertheless.

In the first lines of his book *The Open: Man and Animal* (2002), the Italian philosopher Giorgio Agamben writes of a thirteenth-century Jewish Bible in the collection of the Ambrosian Library in Milan, a Bible 'that contains precious miniatures'. He refers us in particular to its last two pages, in the centre of the first of which (depicting the vision of Ezekiel) 'are the seven heavens, the moon, the sun, and the stars', and in the corners of which 'are the four eschatological animals: the cock, the eagle, the ox, and the lion'.

The last page, he tells us, is divided into halves, the upper of which represents 'the three primeval animals': the bird (Ziz, 'in the form of a winged griffin'), the ox (Behemoth), and the fish (Leviathan) 'immersed in the sea and coiled upon itself'. 'The scene that interests us in particular here', he then says:

> is the last in every sense, since it concludes the codex as well as the history of humanity. It represents the messianic banquet of the righteous on the last day ... [C]heered by the music of two players, the righteous, with crowned heads, sit at a richly laid table. The idea that in the days of the Messiah the righteous, who for their entire lives have observed the prescriptions of the Torah, will feast on the meat of Leviathan and Behemoth without worrying whether their slaughter has been kosher or not is perfectly familiar to the rabbinic tradition. What is surprising, however, is one detail that we have not yet mentioned: beneath the crowns, the miniaturist has represented the righteous not with human faces, but with unmistakably animal heads. Here, not only do we recognize the eschatological animals in the three figures on the

right – the eagle's fierce beak, the red head of the ox, and the lion's head – but the other two righteous ones in the image also display the grotesque features of an ass and the profile of a leopard. And in turn the two musicians have animal heads as well – in particular the more visible one on the right, who plays a kind of fiddle and shows an inspired monkey's face. (1–2)

This is the first of two visions Agamben offers of what we might call, with him, the eschatological animal. The second comes via the philosopher Georges Bataille's fascination with some Gnostic effigies 'of animal-headed archons' that he saw in the Bibliothèque Nationale in 1930 and that had a significant influence upon his subsequent thinking.

'In Gnostic mythology,' Agamben explains, 'the archons are the demonic entities who create and govern the material world, in which the bright and spiritual elements are found mixed and imprisoned [standard Manichean doctrine] in those dark and bodily' (5). One of the effigies concerned – they were in fact medals – presents 'three archons with duck heads', another a 'god with the legs of a man, the body of a serpent, and the head of a cock', another an 'acephalous god topped with two animal heads'.[2] Six years later Bataille and others published the first issue of a journal named *Acéphale* ('headless'), the 'programmatic text' (Agamben's term) of which read 'Man has escaped from his head, as the condemned man from prison' (*L'homme a échappé à sa tête comme le condamné à la prison*).

There are a couple of things to say here. Firstly that another significant influence upon Bataille was Alexandre Kojève, whose lectures on Hegel Bataille attended at the École des Hautes Études between 1933 and 1939, and that in some large part these lectures focused upon the problem of the end of history and the forms human and non-human animals would take in the world beyond it. And secondly, that the 'programmatic text' of *Acéphale* quoted above responds quite directly to a statement by Nietzsche in *Human, All Too Human* (1878), that 'We behold all things through the human head and cannot cut off this head'.[3]

---

2   Bataille's own descriptions. How one can be at once *acephalous* and be supplied with two heads I don't quite see, but no matter.

Two visions, then, of (non-human) animals at the end of time, one as the righteous, invited to sup – and perhaps to judge – at the apocalyptic/messianic table, another as 'demonic entities' governing the material world. These characterisations are not all that different from one another, and perhaps we need not see them as such.

Nor, as, in referencing the Book of Ezekiel, the first of those visions itself seems to suggest, is it as if we have to rely upon arcana alone. At the very beginning of Ezekiel (chapter 1, verses 1–10) comes the prophet's vision of the divine chariot, driven by 'the likeness of a man', and made, in effect, by the wings of four creatures, one at each corner, each one of whom has the body of a man (but for the soles of his feet which are like the soles of a calf's foot), and, most significantly, a head which has four faces: one of a man, one of a lion, one of an ox, and one of an eagle.

Traditionally – a tendency to avoid seeing non-human animals even when they are directly before us – the animals whose faces these are have been effectually effaced, if not occluded, by the human qualities those animals are seen to symbolise: the power of the lion, the servitude of the ox, and the vision of the eagle. Anything, one feels, to avoid the gist most clearly before us, that to see the earth as God sees it we must view it not merely through human eyes but through the eyes of the other creatures we share it with – that, and that these animals, the human animal but one amongst them, are not the guardians or haulers of the chariot, they *are* the chariot.

But how to figure such visions – how *might they figure* – in our thinking, at this point in time?

Since the figure of Nietzsche has entered the discussion, perhaps we might return, briefly, to 1889. (One step forward, one step back: at the end of history such oscillations shouldn't much concern us.)

In Turin, on the 3rd of January of that year – to re-tell a well-known tale – Nietzsche was staying in a house overlooking the Piazza Carlo Alberto. Apparently, on that day, a Thursday, he saw a cab-man cruelly flogging a horse at the other end of the piazza and, running to protect the animal, weeping, threw his arms around his or her neck. He then fell to the ground with what some have speculated was a stroke, others

---

3    Trans. Hollingdale (1996), §9(15).

have claimed was something related to tertiary syphilis, and has generally been accepted as the onset of a madness that was with him until the end of his life almost twelve years later.

In the version of this story that *I* first heard – was it my imagination? – Nietzsche, before falling to the ground, *said* something to the horse, to comfort it. If this were true, then it could be seen as a very significant utterance indeed, particularly if we accept those accounts by which Nietzsche only ever spoke once again (the words 'Mother, I am stupid!' as he came out of his coma two days later).

Much of this is disputed. According to other accounts, for example, Nietzsche, later in his madness, was given to repeating obsessively 'I do not like horses'. More to the point, there appears to be no real evidence that Nietzsche actually saw a horse being mistreated at all. The one surviving account of the incident – by Nietzsche's landlord at the time, although given to a newspaper over a decade later – has it only that, as the landlord was walking on a street nearby, he heard a commotion and, going over, saw Nietzsche in the company of two constables, one of whom told him they had found the philosopher clinging to the neck of a horse. No mention of weeping, no mention of words spoken *to* the horse, no mention of mistreatment or of the philosopher's falling to the ground. These things, some have suggested, may have slipped into the legend via Nietzsche's preoccupation with a passage in Dostoevsky's *Crime and Punishment* in which, just before he commits the murders upon which the book turns, Raskolnikov dreams of himself as a child trying to protect a horse from a vicious flogging – a preoccupation for which there is some corroborating evidence.

Eight months before the incident in Piazza Carlo Alberto, Nietzsche had written to his friend Reinhart von Seydlitz that he had just 'thought' (invented? imagined?) a picture: 'A winter landscape and in the middle an old coachman who, with a cynical expression on his face harder than the surrounding winter, is relieving nature against his horse's legs. The horse, a poor oppressed creature, turns round to look, and is grateful, very grateful.'

What to make of this? What might seem at first to be an act of disdain, even cruelty, might in fact be one of consideration, even love. One's urine-stream, after all, is hot, and so might bring a moment's relief or pleasure to one's suffering horse. The jury is out. In *The Twilight of the Idols*, a work completed just before his collapse and released shortly afterwards, Nietzsche had written, as the eleventh of its 'Maxims and Arrows', 'Can an *ass* be tragic? To be crushed by a burden one can neither bear nor throw off? ... The case of the philosopher.'

The story, that is, may be legend, or may be true. We simply don't know. And perhaps it doesn't matter. Even if, as some have implied, Nietzsche, in his incipient 'madness', staged the incident in the Piazza Carlo Alberto, what would that mean, other than that *he had something to say*, either by/through the scenario, or to the horse him/herself? Certainly it would not alter the two most intriguing questions the incident leaves us with: What, if he spoke at all, *did* Nietzsche say to the horse? and What, after this incident, *became* of that horse?

As to the first, we are not entirely without leads. While it would be a radical misrepresentation to suggest that Nietzsche was a philosopher of 'the animal', or even that, mistreated horses aside, he was particularly preoccupied by non-human animals, he nevertheless turns with some frequency towards them, and on occasions with a certain declarative force. 'The deeper minds of all ages have had pity for animals', he writes in an early essay on Schopenhauer,[4] and his own mind appears to have been one of them. 'I fear the animals regard man as a being like themselves', he writes in *The Gay Science* (1882), 'that has lost in a most dangerous way its sound animal common sense; they consider him the insane animal, the laughing animal, the weeping animal, the miserable animal.' 'Error', he writes most intriguingly in *Human, all too Human* (1878), 'has transformed animals into men; is truth perhaps capable of changing man back into an animal?'[5]

There is also the matter of his early and emphatic vegetarianism. Cosima Wagner writes in her diary entry for 19 September 1869, of her husband Richard's annoyance at his twenty-five-year-old house-guest Nietzsche's refusal to eat meat,[6] a refusal which appears to have changed

---

4   'Schopenhauer as Educator' (1873).
5   §224 (211) and §519 (182) respectively.
6   Skelton (1977): 'Coffee with Prof. Nietzsche; unfortunately he vexes R. very much with an oath he has sworn not to eat meat, but only vegetables. R. considers this nonsense, arrogance as well, and when the Prof. says it is morally important not to eat animals, etc., R. replies that our whole existence is a compromise, which we can only expiate by producing some good. One

only in 1875 when doctors advised the eating of meat was the only way to overcome an unpleasant digestive disorder (Nietzsche writes subsequently of his *weariness* at eating so much meat).

Milan Kundera, in *The Unbearable Lightness of Being*, speculates that when Nietzsche 'put his arms around the horse's neck and burst into tears', he 'was trying to apologize to the horse of Descartes. His lunacy (that is, his final break with mankind) began at the very moment he burst into tears over the horse.' The incident seems also to have long interested Bela Tarr, the Hungarian director of, amongst other auteur masterpieces, *Satantango*, at seven hours one of the longest films I've ever encountered. Tarr says he first heard the story of Nietzsche and the horse from his long-time collaborator/scriptwriter, Lásló Krasznahorkai, in the mid-1980s, and that, as for many others (myself included), his own first response was 'What happened to the horse?' His 2011 film *The Turin Horse* can't be said to answer that – who could? – but offers an intriguing scenario. There has been a teleology to Tarr's oeuvre, each bleak, existential film a development upon the previous. He says – and it is clear enough from the film itself – that *The Turin Horse* is about the end of time. He also says that it is the last film that he will make.

Arguably, the question as to what happened to the horse is a fanciful one in the first place. Given (if the story is true) that the horse was being cruelly beaten by the cab-man, we can assume that it – but no horse is an *it* – would not have been lovingly buried when he or she passed peacefully away, and that a more likely scenario is that at some point, no longer of use (he or she was probably being beaten because he or she was already refusing to budge), the horse was, as are the majority of failed/spent horses to this day (most of them racehorses), taken to the knackery to be slaughtered, flayed, sectioned, his or her flesh separated to be fed to other animals, bones ground for fertiliser, and hooves and joints boiled down for glue and other by-products. For the horse in question, however, this might still have been some time off. A few years, a few months, a few weeks. Perhaps this is what Bela Tarr has in mind.

---

cannot do that just by drinking milk – better, then, to become an ascetic. To do good in our climate we need good nourishment' (148).

Tarr films in black and white, so expertly that this absence of colour becomes in itself a broad spectrum, and specialises in long takes: there are only thirty in *The Turin Horse*, an hour and twenty-six minute film. The opening take, eight minutes long, is a piece of extraordinary cinematic power. In a winter landscape, we see the horse, lashed repeatedly by a man who appears no less troubled than his victim, pulling a cart into a strong headwind, his (her?) blinkered face full of effort and anguish as dead leaves and bits of debris blow into and about it.

If this is, as the title clearly leads us to believe, the horse that was beaten in the piazza in Turin, it is now in another place entirely. Perhaps, at the end of the world, we have also come to the end of geography as we've known it. It is not so much that it would be hard to imagine any place near Turin quite so bleak as this, but the particular, bare, impoverished rural landscape is at once virtually indistinguishable from that undisguisedly Hungarian landscape of several of Tarr's previous films (*Damnation, Satantango, The Werckmeister Harmonies*), and a landscape which, even in this dramatic opening scene, owes as much to Samuel Beckett as it does to any place we might find on Google Earth.

From Turin in 1889, that is to say, we have somehow arrived in a harsh, eschatological waste land. It may be 1889; it may be 1952; it may be 2002. It's not, after all, as if this film is concerned to avoid inconsistency; it might even be making a point of it. The cab-man, Ohlsdorfer, lives with his daughter in relative poverty. Every day she must carry water from a well some thirty or forty metres from the cottage. Every day (for example) they appear to eat only one potato each, boiled whole, in the evening. Starving as they must be, however, and in a film which is hardly going to deny them the time to do so, neither of them (we see this ritual repeatedly) *finish* their potato. Instead – absurdly for such poor people – they eat only a few bites before the daughter takes the plates away, scrapes the unfinished potato into the bin.

But I get ahead of myself. When, in the second take of the film, the cab-man (become cart-man) arrives home, his daughter helps him to free the horse, a mare, from the cart and unharness her. She is then led into her stable. The wind is howling the whole time, though what in fact we hear is a sound-loop, the same few bars of wind-sound played over and over. We do not see the horse again until, thirty minutes of film-time later, she is brought out the next morning and harnessed to the cart. When Ohlsdorfer signals her to move, however, she will not budge. Ohlsdorfer beats her but soon relents and she is returned to the stable. Thereafter we see her only another three times, twice when the daughter tries to get her to eat – but she will not eat, nor drink – and the stable is mucked out, and once when Ohlsdorfer, making a failed attempt to leave the place, leads the horse (it is the daughter who pulls the laden cart) over the horizon, only – a very long take – to reappear on it and return home, as if what they have found is that the rest of the world has ceased to exist.

Meanwhile they have had a visitor, a neighbour who has come to buy brandy ('Why didn't you go to town for it? Ohlsdorfer asks. 'Because the town is no longer there,' the neighbour replies), and who, in the only speech in the film longer than a brief sentence or two, gives an account of the way the world is ending owing to the rapaciousness of humankind, the delinquency of the aristocracy, and the malevolence of God. I thought of Pozzo in *Waiting for Godot*, and of him again when, across this bleak, virtually treeless landscape, suddenly (and yet gradually: another long take), a cart-load of gypsies arrives and there is an altercation as first the daughter and then Ohlsdorfer himself try to repel them and deny them water from the well. 'This is for the water,' one of the Romany says as they leave, handing the daughter a book.

The next day the well is dry. A punishment? A curse? A spell? The daughter sits down to read the book. It is a kind of anti-Bible (Tarr's own description). 'Since in holy places,' it tells her,

> only those things are allowed the practice of which serves the veneration of the Lord, and everything is forbidden that is not fitted for the holiness of the place, and since holy places have been violated by the great injustice of actions that have taken place within them that scandalize the congregation, for this very same reason no service unto the Lord can be held there until, through a ceremony of penitence, these aforementioned injustices have been put to rights. The bishop says to the congregation, 'The Lord was with you! Morning will become night, night will be at an end.'

The winding down of the world continues. Through a fourth, fifth and last, sixth day. As if to reverse the seven days of creation, leaving us one day short, like an open question. On the fifth day the fire goes out and cannot be re-lit. On the sixth the world is plunged into darkness in the mid-morning. The lamp will not light. Oldsdorfer and his daughter sit, resigned, in unnatural night.

*Has* this film been about the horse? One feels that, yet again in a human artefact, a non-human animal has been instrumentalised, to echo and/or underpin the state of humankind (when the horse will not pull the cart it is the daughter who becomes the horse; the pair eat their potatoes 'with their hands like animals';[7] Ohlsdorfer, eventually, declines to eat, much as the horse has done; a reviewer for the *New York Times* refers to Tarr's universe as a place 'indifferent ... to the striving of human beings and other dumb animals'). Instrumentalised, that is, and shut away. *If* this *is* a film about the Turin horse, then some will find it strange she is so much of the time out of sight and mind, given scarcely five per cent of the screen time. But there may be other ways of considering her. Perhaps it is Tarr's *point* that she's been shut away, neglected, ignored. And perhaps there is more.

To her refusal, for example. It seems we have begun at last to realise that animals can commit suicide. And zoo animals, farm animals, working animals, confined and disempowered, will very likely find that their only way to do so is to stop eating – to *refuse* to eat even when pressed to do so. Could this be what the 'Turin' horse has chosen to do? In despair? In protest? In retribution? It is her *refusal*, after all, that, breaking Ohlsdorfer's routine and depriving him of his income, tips this microcosm at last towards its decreation.

He may not have presented the suicide of a *non-human* animal before, but suicide itself has been a focus for Tarr. One thinks particularly of the suicide of the troubled (neglected, ignored) young girl Estike in *Satantango*, a suicide perhaps rehearsed the previous day when she brutalises and then poisons her to all appearances beloved cat (Tarr, one reflects sadly, is not beyond cruelty to non-human animals for the sake of cinema), as if to ensure that the cat comes with her.

---

7    *Filmmaker Magazine*, https://bit.ly/32tMLiw.

But I was speaking of other ways. In *Satantango* the demonically charismatic Irimiás, thought dead a year and a half, is welcomed back into the village but brings with him a plan that will empty and destroy it. In what, chronologically, might in fact be its *final* scene, the film opens with a very long and haunting take of a herd of cows moving slowly, at dusk, amongst abandoned houses. Archons? 'If history is nothing but the patient dialectical work of negation,' writes Agamben in another memorable passage,

> and man both the subject and the stakes in this negating action, then the completion of history necessarily entails the end of man, and the face of the wise man who, on the threshold of time, contemplates this end with satisfaction necessarily fades, as in the miniature in the Ambrosian, into an animal snout. (7)

An echo tracks us through Tarr's last film and it may behove us, before we attempt anything like a conclusion, to face it. The *Turin* horse, and the *Trojan* horse: could there be anything to it? The resonance here is so strong we could almost think it hidden in plain sight. Might it be that, in bringing her back from her encounter with Nietzsche in the Piazza Carlo Alberto and taking her, as it were, into his house (a very normal thing to do – their routine, after all – but she has changed), Ohlsdorfer has also brought into his house *something she now carries with her*, something that occurred – that entered or attends her – through her transaction with the philosopher, something that, as was the case with the Trojan horse, will bring about a judgement upon and the consequent destruction of that house?

Whether or not this is something Nietzsche said is another matter. *Refuse? Rebel? The End has arrived? The time has come?* What *could* he have uttered, if anything at all, to give her such resolve, to turn her, if only as avatar, into one of the horses of revelation, one of the animal-faced archons, passing its ruling at the end of time?

And what if, anthropocentric, *linguocentric*, this is the wrong approach, the wrong question entirely? What if the real question is not what Nietzsche said to the horse, but what the horse said to Nietzsche, or rather, since that would have been nothing, given that horses can't speak in any language that human animals can readily understand, *what that* Nothing *said*, what it *exposed*, to drive Nietzsche mad, or send him so darkly sane: to take away *his* power of speech?

I like to think of it, this Nietzschean moment – this Nothing – as a sudden encounter, at once epiphanic and utterly disgrounding, with, on the one hand, the force and intensity of another animal's *being*, and on the other a yawning gap, a terrible, concomitant blindness, a profound guilt in oneself, as if some substrate of one's own being (an assumption of its rightness and centrality) has been abruptly removed, the apprehension of the other's being entailing, as it does, a realisation of our complicity in the (ab)use of and mindless cruelty towards that other, and the way this in its turn has meant the abandonment of some key thing in our *own* being – an existential *validity* – without which we are, and have, nothing.

FROM ANOTHER PERSPECTIVE, of course – if we are to believe any of our recent speculations concerning the age of the universe, etc. – there is no end of time. The only *actual* end of time that any of us can know is personal, individual. Each of us – if we are given any warning, any chance – must deal with our own private apocalypse. For some this will be no large matter, simply a running out of time as we have, individually, been knowing it. For others it may be of paramount concern. We can go to the grave as unthinking, unrepentant abusers of other creatures we share the planet with; we can go with no real thought for such creatures; or we can go with regret for our role in their suffering, and concern to do something to relieve that suffering. There are no imperatives, and it may come down, for the individual, to a matter of conscience, a matter of belief.

Do *I* really think that 'the animals' will judge us? Yes and no. I think it more likely they have judged us already. What is the impact of that judgement? I don't know. A kind of inner rot, perhaps. It may not be about the end of time at all. The archons may not be telling us of a judgement to come, they may be speaking of a judgement already made, by ourselves, upon ourselves (cum Nietzsche as just hypothesised), a shame, consciousness of a crime – an error – so deeply suppressed that we have, most of us, only its impact, without any sense of the source of that impact. If, from our dreams, our artefacts, the endpapers of our sacred books, the archons are offering us anything at all, it may be, on the one hand, a chance to realise or re-apprise that source, and our own consequent woundedness, and, on the other, a way, before the ends of our own times, to salve it in the long and difficult work of redress.

# Works Cited

Ackland, Michael (1994). *That Shining Band: A Study of Australian Colonial Verse Tradition*. St Lucia: University of Queensland Press.

Adorno, Theodor (1978 [1951]). *Minima Moralia: Reflections from Damaged Life*. E.F.N. Jephcott, trans. London: Verso Editions.

Agamben, Giorgio (2004 [2002]). *The Open: Man and Animal*. Kevin Atell, trans. Stanford: Stanford University Press.

Appelbaum, David (2009). *Jacques Derrida's Ghost: A Conjuration*. Albany, NY: SUNY Press.

Auty, John (2005). Red Plague, Grey Plague. In *Kangaroos: Myths and Realities*. Maryland Wilson and David B. Croft, eds. 56–62. Melbourne: Australian Wildlife Protection Council.

Barthes, Roland (1981). *Camera Lucida*. Richard Howard, trans. New York: Hill & Wang.

— (1977 [1968]). The Death of the Author. In *Image, Music, Text*. Stephen Heath, ed. and trans. London: Fontana Press.

Bataille, Georges (1970). *Oeuvres completes*, vol. II. Paris: Gallimard.

— (1936). La Conjuration Sacré. *Acéphale* 1. 24 June.

Baudelaire, Charles (1982). *Les Fleurs du Mal*. Richard Howard, trans. Jaffrey NH: David R. Godine.

Benfield, W.F. (2015). *At War with Nature: Corporate Conservation and the Industry of Extinction*. Kindle edition. Wellington, NZ: Tross Publishing.

— (2015). The Lord Howe Island Rat Eradication Program. *Tasmanian Times*. 4 June.

Bezan, Sarah, and James Tink, eds (2017). *Seeing Animals: Derrida, Visuality, and Exposures of the Human*. New York: Lexington Books/Rowman & Littlefield.

Blanchot, Maurice, and Jacques Derrida (2000). *The Moment of My Death/Demeure*. Elisabeth Rottenberg, trans. Stanford: Stanford University Press.

— (1999 [1941]). *Thomas the Obscure*. Robert Lamberton, trans. In *The Station Hill Blanchot Reader*. George Quasha, ed. 51–128. Barrytown, NY: Station Hill Press.

— (1995 [1947]). Literature and the Right to Death. In *The Work of Fire*. Charlotte Mandell, trans. Stanford: Stanford University Press.

— (1986 [1980]). Our Clandestine Companion. In *Face to Face with Levinas*. Richard A. Cohen, ed. Albany NY: SUNY Press.

— (1981 [1955]). The Gaze of Orpheus. In *The Gaze of Orpheus, and Other Literary Essays*. Lydia Davis, trans. Barrytown, NY: Station Hill Press.

Bourdain, Anthony (2000). *Kitchen Confidential: Adventures in the Culinary Underbelly*. New York: Bloomsbury.

Breton, André (1969 [1935]). Political Position of Today's Art. In *Manifestoes of Surrealism*. Richard Seaver and Helen R. Lane, trans. 212–33. Ann Arbor: University of Michigan Press.

Brooks, David (2019). *The Grass Library*. Blackheath, NSW: Brandl & Schlesinger.

— (2018). Curator. *Kangaroos: 100 Days Project*. https://bit.ly/2VN7VUB.

— and Michele Hamadache (2017). For the Animals. Interview. *Meanjin* 76.2: 162–70.

— (2016). *Derrida's Breakfast: Poetry, Philosophy, Animals*. Blackheath, NSW: Brandl & Schlesinger.

— (2016). Roogate. *Bungendore Bulletin*, May: 7–10.

— (2007). Charles Harpur and the Warp: Strange Happenings in 'The Creek of the Four Graves'. *Southerly* 67(1/2): 151–56.

— (1999). Scheherazade, the Search for Story: Recent Australian Fiction set in China and Southeast Asia. *Southerly* 59(3/4): 228–38.

— (1985). The Book of Sei. In *The Book of Sei and Other Stories*. Sydney: Hale and Iremonger.

Browne, Thomas (1956 [c1650]). On Dreams. In *Religio Medici and Other Works*. New York: Gateway Editions Inc.

Byrnes, John V. (1961). Barron Field – Recultivated. *Southerly* 21(3): 6–18.

Calarco, Matthew (2008). *Zoographies: the Question of the Animal from Heidegger to Derrida*. New York: Columbia University Press.

Callicott, J. Baird (1980). Animal Liberation: A Triangular Affair. *Environmental Ethics* 4: 311–38.

Cixous, Hélène (2005 [1998]). *Stigmata*. Eric Prenowitz et al, trans. London: Routledge.

Clendinnen, Inga (2007). About Bones. In *Agamemnon's Kiss*. Melbourne: Text Publishing.

Cousins, A.D. (1999). Barron Field and the Translation of Romanticism to Colonial Australia. *Southerly* 58(4): 157–74.

D'Alpuget, Blanche (1981). *Turtle Beach*. Harmondsworth: Penguin Books.

Darwin, Charles (1965 [1872]). *The Expression of the Emotions in Man and Animals*. Chicago: University of Chicago Press.

Derrida, Jacques (2009 [vol. I] and 2011 [vol. II]). *The Beast & The Sovereign*. Michel Lisse, Marie-Louise Mallet and Ginette Michaud eds, Geoffrey Bennington trans. Chicago: University of Chicago Press.

— (2008). *The Animal That Therefore I Am*. David Wills, trans. New York: Fordham University Press.

— and Elisabeth Roudinesco (2004). *For What Tomorrow … : A Dialogue*. Jeff Fort, trans. Stanford: Stanford University Press.

— (2002). The Animal That Therefore I Am. David Wills, trans. *Critical Inquiry* 28(2): 369–418.

— (1993a). *Khora*. Paris: Galilee.

— (1993b). Circumfession. In Geoffrey Bennington and Jacques Derrida. *Jacques Derrida* . Chicago: University of Chicago Press.

— (1986 [1974]). *Glas*. John P. Leavey Jr. and Richard Rand, trans. Lincoln and London: University of Nebraska Press.

— (1976). *Of Grammatology*. Gayatri Chakravorty Spivak, trans. Baltimore: Johns Hopkins University Press.

Eagleton, Terry (1983). *Literary Theory: An Introduction*. London: Blackwell.

Eason, C.T., J. Ross and A. Miller (2013). Secondary poisoning risks from 1080-poisoned carcasses and risk of trophic transfer: a review. *New Zealand Journal of Zoology* 40(3): 217–25. DOI: 10.1080/03014223.2012.740488.

Emery, Lea (2016). The ugly truth about that 'grieving roo' photograph. *Fraser Coast Chronicle*, 14 January. https://bit.ly/2JUPbQq.

Feltman, Rachel (2016). Photos of 'grieving' kangaroo actually show necrophilia (and possibly a killing). *Washington Post*, 15 January. https://wapo.st/39I49Em.

Fetterley, Judith (1978). *The Resisting Reader*. Bloomington: Indiana University Press.

Formosa, Amy (2016). Photographer captures kangaroo family's grief. *Fraser Coast Chronicle*, 13 January. https://bit.ly/3gk73jV.

Frazer, James George (1963 [1922]). *The Golden Bough: A Study in Magic and Religion*. Abridged edition. London: Macmillan & Co.

Freud, Sigmund (1982 [1929]). *Civilisation and Its Discontents*. Joan Riviere, trans. London: Hogarth Press and the Institute of Psycho-analysis.

— (1976 [1900]). *The Interpretation of Dreams*. James Strachey, trans. The Pelican Freud Library: Volume 4. Harmondsworth: Penguin Books.

Fukuyama, Francis (1992). *The End of History and the Last Man*. New York: Free Press.

Genet, Jean (1988 [1958]). *What Remains of a Rembrandt Torn into Four Equal Pieces and Flushed Down the Toilet*. Bernard Frechtman and Randolph Hough, trans. Madras and New York: Hanuman.

Gray, Robert and Geoffrey Lehmann, eds (2011). *Australian Poetry since 1788*. Sydney: University of New South Wales Press.

Grenville, Kate (2005). *The Secret River*. Melbourne: Text Publishing.

Haraway, Donna (2008). *When Species Meet*. Minneapolis and London: University of Minnesota Press.

— (1991). *Simians, Cyborgs, and Women: The Reinvention of Nature*. New York: Routledge.

Heidegger, Martin (1995 [1983]). *The Fundamental Concepts of Metaphysics*. William McNeill and Nicholas Walker, trans. Bloomington: Indiana University Press.

— (1971). *Poetry, Language, Thought*. Albert Hofstadter, trans. New York: Harper and Row.

Heinrich, Joseph, Steven J. Heine and Ara Norenzayan (2010). The Weirdest People in the World? *Behavioural and Brain Sciences*, 33(2/3): 61–83.

Hindwood, K.A. (1938). The Extinct Birds of Lord Howe Island. *Australian Museum Magazine* 6(9): 319–24.

Hunt, Elle (2016). Kangaroo in 'grieving' photos may have killed while trying to mate, scientist says. *Guardian*, 14 January. https://bit.ly/3qyTLVq.

Jakobson, Roman, and Claude Levi-Strauss (1987 [1962]). Baudelaire's 'Les Chats'. In Roman Jakobson, *Language in Literature*. Krystyna Pomorska and Stephen Rudy, eds. 180–97. Cambridge, Mass.: Harvard University Press.

— (1972 [1960]). Linguistics and Poetics. In *The Structuralists from Marx to Lévi-Strauss*. Richard and Fernande De George, eds. New York: Anchor Books.

Jones, Gail (2008). *Sorry*. Melbourne: Penguin Books.

Jones, Robert (1889). Vegetarianism, with Special Reference to its Connection with Temperance in Drinking. Lecture given at the Total Abstinence Society, Melbourne, 10 April 1888, 2nd edn. Joseph Knight, ed. Melbourne and Manchester: The Vegetarian Society.

Joy, Melanie (2009). *Why We Love Dogs, Eat Pigs, and Wear Cows: An Introduction to Carnism*. Berkeley: Conari Press.

Kane, Paul (1996). *Australian Poetry: Romanticism and Negativity*. Cambridge, New York, Melbourne: Cambridge University Press.

Kramer, Leonie and Adrian Mitchell, eds (1985). *The Oxford Anthology of Australian Literature.* Melbourne: Oxford University Press.

Kristeva, Julia (1984). *Revolution in Poetic Language.* Margaret Waller, trans. New York: Columbia University Press.

Lacan, Jacques (1977 [1949]). The Mirror Stage as Formative of the Function of the I as Revealed in Psychoanalytic Experience. In *Écrits: A Selection.* Alan Sheridan, trans. New York: W.W. Norton & Company.

— (1977 [1957]). The Agency of the Letter in the Unconscious. In *Écrits: A Selection.* Alan Sheridan, trans. New York: W.W. Norton & Company.

Lawson, Henry (1972). *Short Stories and Sketches, 1888–1922.* Colin Roderick, ed. Sydney: Angus & Robertson.

Legge, S., Murphy, B.P., McGregor, H. et al. (2017). Enumerating a continental-scale threat: how many feral cats are in Australia?, *Biological Conservation* 206 (February), 293–303.

Lewis, Martin W. (2010). Lord Howe Island: Return of the Tree Lobster. *GeoCurrents* 8. https://bit.ly/3qAovFC.

Lifton, Robert Jay (1988). *The Nazi Doctors: Medical Killing and the Psychology of Genocide.* New York: Basic Books.

Lorca, Federico García (1998 [1933]). Play and Theory of the Duende. In *In Search of Duende.* Christopher Maurer, ed. New York: New Directions.

Mathews, Freya (2012). The Anguish of Wildlife Ethics. *New Formations* 76: 114–31.

McCulloch, Allan R. (1921). Lord Howe Island – A Naturalist's Paradise. *Australian Museum Magazine* 1(2): 31–47.

McElligott, Alan G, Kristine H. O'Keefe and Alexandra C. Green (2020). Kangaroos display gazing and alternations during an unsolvable problem task. *Biology Letters* 16.12 (Royal Society). https://doi.org/10.1098/rsbl.2020.0607.

Mears, Gillian (2011). *Foal's Bread.* Crows Nest, NSW: Allen & Unwin.

Metcalf, Peter and Richard Huntington (1979). *Celebrations of Death: the Anthropology of Mortuary Ritual.* London: Cambridge University Press.

Miller, Alex (2007). *Landscape of Farewell.* Crows Nest, NSW: Allen & Unwin.

— (2002). *Journey to the Stone Country.* Crows Nest, NSW: Allen & Unwin.

Moyal, Ann (2001). *Platypus.* Crows Nest, NSW: Allen & Unwin.

Nietzsche, Friedrich (1977). *A Nietzsche Reader.* R.J. Hollingdale, ed. and trans. Harmondsworth: Penguin.

— (1968 [1889]) *Twilight of the Idols.* R.J. Hollingdale, trans. Harmondsworth: Penguin Books.

— (1974 [1882]) *The Gay Science.* Walter Kaufman, trans. New York: Vintage.

— (1996 [1878]) *Human, All Too Human*. R.J. Hollingdale, trans. Cambridge: Cambridge University Press.

Ofrat, Gideon (2001). *The Jewish Derrida*. Peretz Kidron, trans. Syracuse, NY: Syracuse University Press.

Paterson, A.B. ('Banjo') (1942). *The Collected Verse of A.B. Paterson*. Sydney: Angus & Robertson.

Patterson, Charles (2002). *Eternal Treblinka: Our Treatment of Animals and the Holocaust*. New York: Lantern Books.

Pound, Ezra (1973 [1911]). I Gather the Limbs of Osiris. *Selected Prose*. New York: New Directions.

Pribac, Teya Brooks (2021). *Enter the Animal: Cross-Species Perspectives on Grief and Spirituality*. Sydney: Sydney University Press.

Readhead, Harry (2016). No, this kangaroo wasn't grieving – it was raping a dead female. *Metro news*, 14 January: https://bit.ly/3mPQE9m.

Recher, H.F. (1974). Colonisation and Extinction: the Birds of Lord Howe Island. *Australian Natural History* 18(2): 64–68.

Reich, Wilhelm (1970 [1933]). *The Mass Psychology of Fascism*. Vincent R. Carfagno, trans. New York: Farrar, Straus & Giroux.

Richards, I.A. (1929). *Practical Criticism*. London: Kegan Paul.

Rilke, Rainer Maria (2011). *Selected Poems*. Robert Vilain, ed. Susan Ranson and Marielle Sutherland, trans. London: Oxford University Press.

— (2009). *Duino Elegies and the Sonnets to Orpheus*. Stephen Mitchell, ed. and trans. New York: Random House.

— (1993). *Duino Elegies and the Sonnets to Orpheus*. Robert Hunter, trans. Eugene, Oregon: Hulogosi.

— (1987a). *The Selected Poetry of Rainer Maria Rilke*. Stephen Mitchell, ed. and trans. London: Picador.

— (1987b). *The Sonnets to Orpheus*. David Young, trans. Middletown, CT: Wesleyan University Press.

— and Katharina Kippenberg (1954). *Briefwechsel*. Bettina von Bomhard, ed. Wiesbaden: Insel Verlag.

— (1949). *The Sonnets to Orpheus*. J.B. Leishman, trans. London: The Hogarth Press.

— (1948). *Letters of Rainer Maria Rilke: 1910–1926*. J.B. Greene and M.D. Herder Norton, trans. New York: W.W. Norton and Co.

— (1942). *The Sonnets to Orpheus*. M.D. Herter Norton, trans. New York: W.W. Norton and Co.

— (n.d.). *The Sonnets to Orpheus* Howard A. Landman, trans. https://bit.ly/2VN7hq9.

— (n.d.). *The Sonnets to Orpheus*. Robert Temple, trans. https://bit.ly/2JVVcwp.

# Works Cited

Ruskin, John (1890 [1856]). *Modern Painters*. London: John Wiley.

Sagoff, Mark (1984). Animal Liberation and Environmental Ethics: Bad Marriage, Quick Divorce. *Osgoode Hall Law Journal* 22(2): 297–307.

Scott, John A. (2014). *N.* Blackheath, NSW: Brandl & Schlesinger.

Sebald, W.G. (1998). *The Rings of Saturn*. Michael Hulse, trans. New York: Vintage.

Seleh, Pardes (2016). 'Grieving' kangaroo photos may actually show brutal murder scene. *Daily Wire*, 15 January. https://bit.ly/3mRsBXp.

Singer, Isaac Bashevis (1968). The Letter Writer. In *The Séance and Other Stories*. New York: Farrar, Straus and Giroux.

Spencer, Colin (1994). *The Heretic's Feast: A History of Vegetarianism*. London: Fourth Estate.

Staker, Lynda (2014). *Macropod Husbandry, Healthcare and Medicinals*, vol. 1. Lynda Staker (publisher).

Stein, Gertrude (1936). *The Geographical History of America, or the Relation of Human Nature to the Human Mind*. New York: Random House.

Stoppard, Tom (1967). *Rosencrantz and Guildenstern Are Dead*. London: Faber & Faber.

Thon, Melanie Rae (2017). Galaxies Beyond Violet. In *After Coetzee: An Anthology of Animal Fictions*. A. Marie Hauser, ed. Minneapolis: Faunary Press.

Tiffin, Helen (2019a). Australian Conservation Policies and the Owls of Lord Howe Island. In *Ecocritical Concerns and the Australian Continent*, Helen Tiffin and Beate Neumeier, eds. New York: Lexington Books.

— (2019b). *Knowing Animals* 120 (podcast): https://bit.ly/39MD1UE.

van Dooren, Thom (2015). A Day with Crows: Rarity, Nativity and the Violent-Care of Conservation. *Animal Studies Journal*, 4(2): 1–28. https://ro.uow.edu.au/asj/vol4/iss2/2.

Wagner, Cosima (1977). *Cosima Wagner's Diaries 1869–1877*. Geoffrey Skelton, trans. New York: Harcourt, Brace, Jovanovich.

Wimsatt, William and Monroe K. Beardsley (1954 [1949]). The Affective Fallacy. In *The Verbal Ikon: Studies in the Meaning of Poetry*. Lexington: University of Kentucky Press.

— (1954 [1946]). The Intentional Fallacy. In *The Verbal Ikon: Studies in the Meaning of Poetry*. Lexington: University of Kentucky Press.

Woodford, James (2010). Rats! Lord Howe's Owls may be Sent Home to Roost. *Sydney Morning Herald*, 23 June. https://bit.ly/39MFXk9.

*Xinhua* (2016). Grieving kangaroo actually just wants sex. *Xinhua*, 15 January. https://on.china.cn/2VLVpow.

# Acknowledgements

The author acknowledges the following prior publications of essays in this volume, and thanks the editors concerned. 'The Smoking Vegetarian' was first published, in a longer version, in *Angelaki: Journal of the Theoretical Humanities*, 14.2 (Oxford: August 2009). This condensed version first appeared in *Best Australian Essays of 2010*, edited by Robert Drewe (Melbourne: Black Inc, 2010). 'Cracks in the Fray' first appeared in the online journal *Jacket* (39) in 2010. 'The Fallacies' first appeared in *Southerly* 73.2 (2013). 'The Loaded Cat', 'Meeting Place' and 'At Duino' were first published in *Derrida's Breakfast* (Blackheath: Brandl & Schlesinger, 2016). In revised and expanded form, 'The Loaded Cat' also appears in *Seeing Animals: Derrida, Visuality, and Exposures of the Human*, ed. Tink and Bezan (New York: Lexington Press, 2017), and 'At Duino' in *Rilke's Sonnets to Orpheus*, ed. Eldridge and Fischer, Philosophical and Critical Perspectives (New York and London: Oxford University Press, 2019). 'Field's Kangaroo' first appeared in the *Kenyon Review* (U.S.) 39.2 (March/April 2017), and 'Writing Animals' in the *Kenyon Review* 41.6 (Nov/Dec 2019). 'Dougald's Goat' first appeared in *The Novels of Alex Miller*, ed. Robert Dixon (Sydney: Allen & Unwin, 2012), and 'The Rats of Lord Howe' online on the Australian Broadcasting Commission's Religion & Ethics page in August 2019. A shorter version of 'An Exoneration' appeared in *Animal Studies Journal* 9.1 (2020), entitled

'The Grieving Kangaroo Photograph Revisited'. All other essays in this volume appear here for the first time.

'The Eighth Elegy' and 'The First Elegy' translation copyright © 1982 by Stephen Mitchell; from *Selected Poetry of Rainer Maria Rilke* by Rainer Maria Rilke, edited and translated by Stephen Mitchell. Used by permission of Random House, an imprint and division of Penguin Random House LLC. All rights reserved.

'Part 2: XI,' and 'Part 2: XIII' from *Duino Elegies & the Sonnets to Orpheus: A Dual-language Edition* by Rainer Maria Rilke, translated by Stephen Mitchell, translation and foreword copyright © 1982, 1985, 2009 by Stephen Mitchell. Used by permission of Vintage Books, an imprint of the Knopf Doubleday Publishing Group, a division of Penguin Random House LLC. All rights reserved.

The author would like to express his gratitude to Melissa Boyde, Fiona Probyn-Rapsey and Yvette Watts, the editors of the Animal Publics series, and to Agata Mrva-Montoya, Denise O'Dea, Jo Butler, and the production team at Sydney University Press.

He would also like to thank, for their expertise, ideas, encouragement, support and inspiration, Andras Berkes-Brandl, Helen Bergen, Ray Drew, Luke Fischer, Steve Garlick, Jason Grossman, Michelle Hamadache, Kevin Hart, John Kinsella, Jeffrey Moussaieff Masson, Ray Mjadwesche, Stella Reid, Scott Stephens, Lynda Stoner, Veronica Sumegi, Evan Switzer, Helen Tiffin, Christine Townend and Veronica Sumegi. Above all, for her strength, rigour, and deep companionship, Teya Brooks Pribac.

Several of these essays were written during the author's tenure as the 2015/16 Australia Council Fellow in Literature, and his time as Associate Professor and subsequently Honorary Associate Professor in Australian Literature at the University of Sydney. He extends deep thanks to each institution for its support.

# Index

www.ingramcontent.com/pod-product-compliance
Lightning Source LLC
Chambersburg PA
CBHW031338020726
47499CB00005B/1320